D1006757

The Climate Fix

Also by Roger Pielke, Jr.

The Honest Broker

The Climate Fix

What Scientists and Politicians Won't Tell You About Global Warming

Roger Pielke, Jr.

BASIC BOOKS
A Member of the Perseus Books Group
New York

Published by Basic Books,
A Member of the Perseus Books Group

Unless otherwise indicated, all illustrations in the book were designed by and reprinted with the permission of Ami Nacu-Schmidt.

Designed by Brent Wilcox

Library of Congress Cataloging-in-Publication Data
Pielke, Roger A., 1968–
The climate fix : what scientists and politicians won't tell you about global warming / Roger Pielke, Jr.
p. cm.
Includes bibliographical references and index.
ISBN 978-0-465-02052-2 (alk. paper)
1. Global warming—Political aspects. 2. Climate change mitigation—Political aspects. 3. Climatic changes—Political aspects. I. Title.
QC981.8.G56P535 2010
363.738'74—dc22

2010021776

10 9 8 7 6 5 4 3 2 1

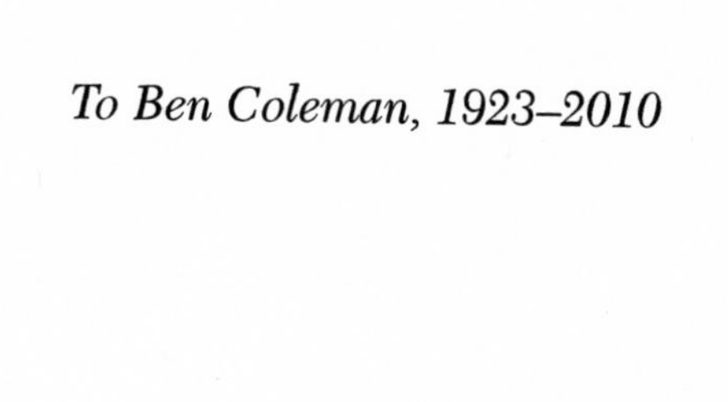

To Ben Coleman, 1923–2010

Contents

Preface

In the fall of 2009 climate policy seemed on track. Expectations for the coming United Nations conference in Copenhagen were measured, but generally optimistic. The optimism seemed warranted. After all, in the previous year the United States—long seen as the single obstacle to coordinated global action on climate change—saw President Barack Obama swept into office, promising in his inaugural address to make responding to climate change a priority of his administration, marking a stark contrast with his predecessor. Soon after, the substantial Democratic majority in the House of Representatives passed "cap and trade" legislation that would put a price on carbon and aimed at reducing U.S. emissions of carbon dioxide by 17 percent by 2020. In the Senate, the Democrats held a supermajority of sixty votes, making prospects for enactment of the legislation highly likely. Leaders in other countries that had been seen as laggards—including Australia, Japan, and even China and India—were saying all the right things about the need for action, joining those in Europe who had long been advocating global action on climate change. Success, it seemed, was finally at hand.

It was remarkable, then, to see the complete meltdown of global climate policy at Copenhagen and the disarray that followed. Far from reaching a truly international agreement, at Copenhagen just a few countries hastily agreed to an accord in the conference's waning hours. Europe was not present in the final negotiations, and the accord itself was merely "taken note of"—diplomatic speak for failure to reach a more substantive agreement. In the United States some environmentalists tried to put on a brave face regarding Copenhagen, but the fact that the agreement was almost universally panned (expect by a few countries interested in business as usual) told the real story. The departure of the UN's chief negotiator—who soon after the conference announced his resignation and intent to work in the private sector—completed the sense of utter failure that was Copenhagen.

The United States saw the momentum for action to cap carbon emissions halt as political winds changed dramatically. The Democrats lost their supermajority in a special election to replace the late Edward Kennedy, senator from Massachusetts. President Obama's decision to prioritize health care reform consumed much of his political capital in the first months of 2010. There seemed little appetite to replicate that effort, and, not surprisingly, ambitious plans for climate legislation were scaled back even before health care reform was enacted, with proposals focusing as much on expanding fossil fuel supply as reining it in. The prospects for major U.S. legislation seemed as far off as a comprehensive global treaty.

As if this was not bad enough, in November 2009 someone stole or released more than a thousand e-mails from a server at the University of East Anglia in the United Kingdom that showed private discussions among climate scientists going back more than a decade. Some of these discussions showed scientists in a rather poor light. Soon thereafter, the Intergovernmental Panel on Climate Change (IPCC) faced criticism after an obvious error was identified in its 2007 report. It didn't help that its initial reaction was to stonewall and deny. A series of further revelations showed a series of errors in the report and breakdowns in its review process. Its chairman was accused of having conflicts of interest. A series of investigations and reviews was subsequently initiated to examine the IPCC and the activities of some of the scientists implicated by the released e-mails. Opinion polls in the United States and Europe showed growing doubts about the trustworthiness of climate scientists.

These failures shouldn't have come as a surprise. The difficulties faced in the politics and science of climate change provide compelling evidence that the course that the world has been on for climate policy has created the conditions for policy failure. For some, the lesson is to reload and try again with the same strategies that have gotten us to where we are today. To me, that seems like insanity. It is time to rethink fundamentally our approach to climate change, and this book offers such a rethinking.

But before you proceed, I offer a warning. Over the past ten years at the University of Colorado I have taught a seminar titled Policy, Science, and the Environment. It seeks to introduce first-year graduate students to the messy intersection of science and politics. On my syllabus I have included a cartoon from the series *Calvin and Hobbes*. Calvin, the little boy, explains to Hobbes, his tiger friend, "The more you know, the harder it is to take decisive action. Once you become informed, you start seeing complexities and shades of gray. You realize that nothing is as clear and simple as it first appears." Calvin explains that he has decided not to risk becoming informed, and Hobbes sympathizes: "You're ignorant, but at least you act on it."

Ignorance, as they say, is bliss, because seeing the world in black-and-white is easy and comfortable. Reality, as Calvin tells us, is actually colored in grays. So too it is in the world of climate change. For some, the climate debate is a morality play, with good guys and bad guys, with virtue and reason on one side and evil and corruption on the other. *The Climate Fix* seeks to clarify the climate debate in a way that anyone who can use a bit of addition, multiplication, and common sense can make sense of. If successful, once you read this book, you'll never see the climate debate in the same way again. And if you want to see progress, rather than gridlock and disarray, seeing the climate debate in a new light might be just the thing we need.

CHAPTER 1

Dinner Table Climate Science for Commonsense Climate Policy

IN THE SUMMER OF 1988 global warming first captured the imagination of the American public. In early June of that summer Senator Al Gore (D-TN) organized a congressional hearing to bring attention to the subject, one that he had been focusing on in Congress for more than a decade. The hearing that day was carefully stage-managed to present a bit of political theater, as was later explained by Senator Tim Wirth (D-CO), who served alongside Gore in the Senate and, like Gore, was also interested in the topic of global warming. "We called the Weather Bureau and found out what historically was the hottest day of the summer. Well, it was June 6th or June 9th or whatever it was. So we scheduled the hearing that day, and bingo, it was the hottest day on record in Washington, or close to it. What we did is that we went in the night before and opened all the windows, I will admit, right, so that the air conditioning wasn't working inside the room."[1]

The star witness that day was James Hansen, a NASA scientist who had been studying climate since the 1960s. Hansen had decided that "it was time to stop waffling so much and say that the evidence is pretty strong that the greenhouse effect is here and is affecting our climate." Hansen emphasized three points in his testimony: First, "the earth is warmer in 1988 than at any time in the history of instrumental measurements"; second, "global warming is now large enough that we can ascribe with a high degree of confidence a cause and effect relationship" to the emission of greenhouse gases, primarily carbon dioxide; and third, the consequences are "already large enough to begin to affect the probability of extreme events such as summer heat waves."[2] The hearing's public impact surely must have exceeded even its organizers' expectations, as the temperature in the room and the scorching weather outside resulted in Hansen's testimony receiving wide coverage in the national and international media.

Not long after the hearing, S. Fred Singer, who like Hansen had spent much of his career as a government scientist and bureaucrat working on climate issues, published an op-ed in the *Wall Street Journal* critical of Hansen's testimony and the reception that it had received.[3] Singer, who had previously publicly questioned the science behind ozone depletion, acid rain, and nuclear winter (and who would later question the science associated with smoking policies), asserted that "more research is needed" before any actions are taken to reduce greenhouse gas emissions due to the very large uncertainties that accompanied the issue. The public battle lines had been drawn on a debate that had been emerging in fits and starts for several decades, if not longer.[4]

Looking back many years later, one observer remarked that the 1988 Gore-Hansen hearing "touched off an unprecedented public relations war and media frenzy," marking "the official beginning of the global warming policy debate."[5] What's more, the hearing had all of the elements that would characterize the debate in the following decades. Politicians sought to stage-manage the scientific community to support their political ambitions. Leading scientists willingly played along, enthusiastically lending the authority of science to the political campaign. Opponents of action engaged the political battle through debates over science—primarily by seeking to raise uncertainty (or, perhaps more accurately, by offering a set of competing certainties) as the basis for opposing efforts to regulate or otherwise address ever-increasing amounts of carbon dioxide and other greenhouse gases in the atmosphere—even though they were and would continue to be representing a minority position on the science. The global warming debate was under way, and how it started set the stage for how it would be fought for the next several decades.

Because political battles over climate change have been fought through science since 1988, it is easy to lose sight of the fact that adversaries on either side of that debate have agreed about core aspects of the science since that time. As I'll argue, that core understanding is sufficient to form the basis for a commensense approach to climate policy. Such an approach will recognize that science can alert us to a potential problem and provide some insight about the consequences of different policy choices, but science cannot decide what choices we ultimately make.

A commonsense approach to climate policy will recognize that there are many justifications for addressing the multiple human and nonhuman influences on climate, and their possible consequences, that should have our attention. For example, in the coming chapters I will introduce a technical concept—decarbonization of the global economy—that lies at the core of any effort to address increasing amounts of carbon dioxide in the atmosphere. Decarbonization refers to efforts to reduce the amount of carbon dioxide associated with economic activity, recognizing that sustaining economic growth is a priority around the world. The world has been decarbonizing for more than a century, and there are good reasons to accelerate that process that have nothing to do with climate science. But I am getting ahead of myself.

Mutual Misunderstandings in Science and Politics . . .

When Jim Hansen was testifying before Al Gore in the summer of 1988, I had just finished my sophomore year at the University of Colorado in Boulder. I was a newly employed student assistant in the Atmospheric Chemistry Division of the National Center for Atmospheric Research. My job was fairly typical of a student assistant in a major research facility: to write simple computer programs that would transfer very large amounts of remote sensing information obtained from earth-orbiting satellites from obsolescing reel-to-reel computer tapes to (then) fancy new storage tapes using the Cray Supercomputers that NCAR (pronounced N-CAR) was world famous for running. The reason for transferring the data was to ensure its continued availability to climate scientists so that they could conduct research using the data at some point in the future. Even in the 1980s the massive volume of data collected by remote sensing technologies such as satellites far exceeded the finite resources available to analyze that data, so archival work was an important (if mundane) part of preserving scientific information for possible future research use.

I had learned scientific FORTRAN programming during the previous two summers while working in a similar student assistant role for researchers at Colorado State University, where my father was a professor of atmospheric science. The first scientific paper that I collaborated on was a result of that summer work. That paper reported the results of an investigation of the effects of fairly regular afternoon cloudiness—such as occurs in the summer when thunderstorms build regularly along the Colorado Front Range—on the orientation of fixed solar panels.[6] We asked whether the solar energy collected by the panels would be enhanced if the panels were shifted east to face more directly the morning sun and away from the cloudier afternoon skies. Before doing the research I had thought that the answer was obvious: of course the panels should be shifted toward the sunnier morning skies. However, upon actually doing the math we learned that a solar panel facing due south still collected more sunlight than one shifted to the east, even under conditions of regular afternoon cloudiness. It was a pretty simple study, yet, for me at least, it delivered unexpected results. It was a good lesson that intuition or belief is very often not a good substitute for actually doing the research, especially for seemingly simple questions with seemingly obvious answers. The lesson was to do the math yourself.

NCAR in the 1980s was a special place. It was not an ivory tower, but it was pretty close to being one. It sits on a mesa above Boulder in a spectacular setting, dwarfed by the foothills yet brilliantly designed by I. M. Pei so as to fit into its surroundings. NCAR has been home to some of the world's greatest thinkers on environmental issues. When I was there in the late 1980s it was not uncommon for giants in the field of atmospheric sciences such as Walter Orr Roberts (NCAR's founder), Will Kellogg, Warren Washington, and Mickey Glantz to join student assistants and other research support staff for lunch and conversation in the cafeteria looking out over the plains at the foot of Colorado's Front Range. Other scientists who were at NCAR at the time included Steve Schneider and Kevin Trenberth, both fixtures of the climate debate before and since that time.

Schneider, a prominent voice in debates about climate since the early 1970s, was even an extra in Woody Allen's 1973 movie *Sleeper*, in which NCAR played a cameo role. The movie was set in 2173, and a few other NCAR employees were also cast as extras. Some bearded scientists didn't make the cut as extras, as Woody Allen apparently did not see facial hair as part of his vision of the future.[7] *Sleeper* is part of NCAR lore, which holds that the institution provides a window to the future. A colleague once remarked to me that just about every leading scholar in the atmospheric sciences had come through NCAR at some point in their career for one reason or another. In that company, I had a front-row seat to watch the atmospheric sciences emerge from being an interesting and relevant area of scientific research to occupy a center-stage position in global political debate.

The late 1980s and early 1990s were heady days for the atmospheric sciences community. The issue of ozone depletion of the stratosphere due to chlorofluorocarbons (a human-made industrial chemical used in refrigeration and air conditioners) was the focus of international attention. The Montreal Protocol governing the production of CFCs was signed in 1987 and subsequently strengthened in later years. Also during that period, policies to address "acid rain," resulting from the emissions from power plants, were being discussed in the U.S. Congress as part of amending the Clean Air Act in 1990. Climate change was emerging as an important policy issue, but also one with the promise of considerable new funding for scientific research. The Intergovernmental Panel on Climate Change, which would share the Nobel Peace Prize with Al Gore in 2007, was begun in 1988, and the so-called Earth Summit in Rio de Janeiro took place in 1992. In 1990 the U.S. Congress passed the Global Change Research Act, which sought to create a comprehensive research program to provide useful scientific information to policy makers grappling with decisions about climate change. A few years later I would write my doctoral dissertation on the ability of this research program to support policy making. In short, this period was one that saw the atmospheric sciences take a prominent role in a range of policy issues of national and global importance.

While I worked as a student assistant at NCAR's Atmospheric Chemistry Division I had the opportunity to listen to the scientific staff as they discussed the relationship of science and politics, typically in the context of discussions of ozone depletion and the responses to it that were being debated at national and international levels. An overarching theme of these discussions among the scientists was that if only policy makers better understood science, then the process of policy making would be so much easier.

Armed with this insight, I decided that it would be valuable to gain some expertise in public policy before returning (I had thought) to a career in the physical sciences. I quit my NCAR student assistant job and was accepted into a graduate program of public policy at the University of Colorado. There I worked on a master's thesis in which I evaluated the performance of the space-shuttle program as compared to the initial promises that NASA had made to secure political support for the program. I worked under the direction of Radford Byerly, a physicist who had spent much of the previous decades as a highly respected congressional staffer for the House Committee on Science and Technology. Rad had come to the University of Colorado a few years before in order to direct a center focused on space and geosciences policy. But his tenure did not last long; in 1991, thanks to his rare knowledge of both science and politics, he was called back to Washington, D.C., to serve as the chief of staff of the Science Committee, under its new chairman, Representative George E. Brown, a Democrat from southern California.

Chief of staff of the Science Committee is a pretty plum position in the world of science policy, so Rad accepted the position and moved back to Washington. I was only halfway through my master's program, so it was potentially a great loss for me. I can only surmise that Rad must have felt guilty about leaving me middegree because he offered me a position in his office as an intern. Much like my student assistant work at NCAR, my duties as an intern for the House Science Committee involved doing the mundane, behind-the-scenes work that makes any large institution work. But thanks to Rad it also gave me a front-row seat to watch the political process in action, especially because Rad made every effort to have me sit in his office—like a potted plant, just taking up space—for important closed-door meetings and to have me tag along with him to meetings and events involving members and senior staff that I never could have observed otherwise.

I will never forget the eye-opening, even life-changing, moment when late one afternoon in Rad's office, the senior staff of the committee were discussing the relationship of science and politics following the visit of a highly respected member of the scientific community, who had come to advocate some political course of action. An overarching theme of this conversation among the staff was that if only scientists better understood policy and politics, then the process of policy making would be so much easier.

For me it was an "Aha!" moment. My experience at NCAR taught me that scientists thought that policy makers needed to better understand science, and my brief stint at the House Science Committee taught me that the policy makers thought the scientists needed to better understand policy and politics. This realization set me forth on a career in science and technology policy, studying (and participating in) that messy intersection of science and politics.

Carbon Dioxide Is Important, but Climate Change Involves Much More

Perhaps it is one of the unavoidable side effects of being the son of a world-famous atmospheric scientist, but I have never questioned the climatic importance of human emissions of carbon dioxide; its importance has always been something that was accepted by my father and presented in his work. You could say I gained a pretty in-depth understanding of the atmospheric sciences at the dinner table. So the "controversy" over whether carbon dioxide emissions affect climate is not a subject that holds much interest for me, and looking back over my published writings on climate change since 1994, there is a consistent message that carbon dioxide does indeed have significant climatic effects. Obviously, this view came straight from my father, and is widely (if not universally) shared in the atmospheric sciences community.

For instance, in the mid-1980s, when my own interests lay far from science and policy, focused instead on soccer and girls, my father wrote an annual article on the atmospheric sciences for the *Encyclopaedia Britannica*. In his 1985 article he explained that emissions of carbon dioxide to the atmosphere would cause a net warming of the Earth's surface due to the fact that it and other trace gases "act to reduce the emission of long-wave radiation out into space yet still permit solar radiation to reach the Earth's surface. This mechanism of heat increase is referred to as the greenhouse effect."[8] In 1984 he wrote that the consequences of an enhanced greenhouse effect could be profound: "Unless mitigated by other results of human activities, such as reduced sunlight at the ground due to additions of aerosols to the upper atmosphere, this warming could result in major changes in climate patterns."[9] Mitigation policies typically focus on efforts to limit the accumulation of carbon dioxide in the atmosphere to some upper limit, a challenge described in technical terms as the "stabilization of carbon dioxide concentrations."

Understanding the challenge of stabilizing carbon dioxide levels in the atmosphere at a constant amount is really quite simple.[10] Imagine you have a bathtub that is filling with water (Figure 1.1). The rising water prompts concern that the tub will overflow, flooding your house and causing damage. Fortunately, there is a hole at the bottom of the tub that is allowing water to drain out of the tub. But unfortunately, this will only put off for a short while the overtopping of the bathtub, as the water is draining out at a rate slower than it is filling. To make matters worse, the rate at which the tub is filling is slowly increasing as each minute goes by.

The challenge that you face is to keep the bathtub from overflowing. Based on the filling rate, its rate of increase, and the open drain, the only way that you can prevent an overflow is by reducing the net rate at which water is filling the tub to zero. In other words, for the water level in the tub to become *stabilized* at a fixed level, the water

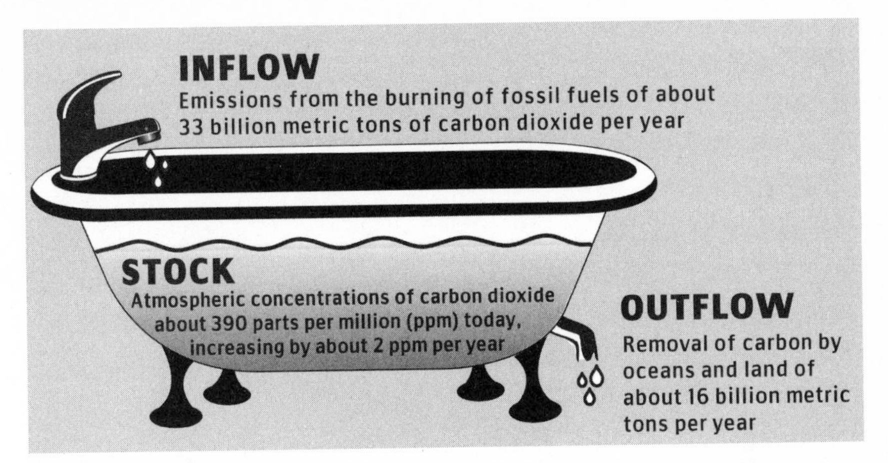

FIGURE 1.1 Understanding the buildup of carbon dioxide.

filling the tub must be coming in at less than or equal to the amount of water being removed.

A simple bathtub model approximates the dynamics associated with the challenge of stabilizing carbon dioxide concentrations in the atmosphere. Human emissions of carbon dioxide (which can be thought of as the water filling the tub from the spigot) are increasing and accumulating in the atmosphere. Scientists measure the amount of carbon dioxide in the atmosphere using the terminology of "parts per million," referring to the amount of carbon dioxide molecules in every million molecules in the atmosphere. At the start of 2010, these values were close to 388 ppm and growing at a rate of about 2 ppm per year during the past decade.

Carbon dioxide accumulates in the atmosphere due to anthropogenic (i.e., human) activities, primarily from the burning of fossil fuels and to a lesser degree from land-use practices such as clearing forests and tilling soil for farming. Several hundred years ago atmosphere concentrations of carbon dioxide were about 280 ppm, meaning that they have increased by more than 100 ppm in the time since.[11] The various human activities that lead to carbon dioxide emissions can be thought of as the spigot in the bathtub analogy from which the water is filling the tub. Atmospheric concentrations have increased because carbon dioxide accumulates faster than it is removed. Emissions are conventionally described in units of billions of metric tons (Mt, about 2,200

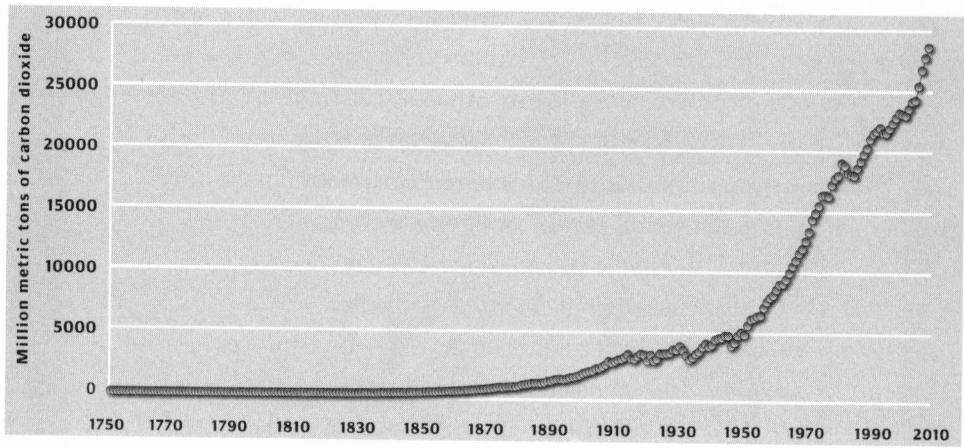

FIGURE 1.2 Carbon dioxide emissions from fossil fuels, 1751–2006. Source: U.S. Department of Energy.

pounds), or gigatons. The addition to the atmosphere of approximately 7.8 Gt of carbon dioxide leads to an increase in CO_2 concentration of 1 ppm.[12] Figure 1.2 shows the increasing emissions of carbon dioxide from the burning of fossil fuels.

Some of the carbon dioxide emitted to the atmosphere through human activity is absorbed by the oceans and by various processes of the land surface.[13] Because the uptake of carbon dioxide is related to processes that change in complex ways due to growing carbon dioxide concentrations, projections of how much carbon dioxide will be taken up by the oceans and land surface in the future are necessarily highly uncertain.[14] The fact that the oceans absorb carbon dioxide reduces the rate at which it accumulates in the atmosphere, which might be good news from the perspective of the atmosphere were it not for the fact that carbon dioxide absorbed into the ocean introduces a different suite of challenges because it leads to changes in the ocean's chemistry, with potentially harmful effects. With accumulating carbon dioxide emissions, unfortunately, the natural system provides no easy, short-term solution.

The ability of the land surface to take up carbon dioxide has been a central feature of international climate policies. If the land surface can store large quantities of carbon dioxide, then this would add some additional time for the global economy to decarbonize, as the amount of carbon dioxide accumulating in the atmosphere would slow down a bit. It is as if the hole in the bottom of the bathtub might be made a bit larger. Such proposals have been particularly appealing to those interested in preserving forests, which store large quantities of carbon (particularly tropical rain forests), as well as farmers, who are able to modulate the amount of carbon stored in their lands through different agricultural practices.

While land-surface management is a potentially valuable short-term tool for sequestering carbon dioxide (as well as for achieving other goals, such as preserving forests, to achieve low atmospheric stabilization targets), it cannot alter the basic need to dramatically accelerate the decarbonization of the global economy. Put another way, whatever is done with respect to land-surface management, it won't change the basic arithmetic of decarbonization.[15] Further, efforts to tie in land-surface management with decarbonization of the economy have been fraught with political challenges. For instance, the state of California found itself under intense criticism from environmental groups when it put forward rules that would allow timber companies to clear-cut old-growth forests and receive financial benefits under a carbon trading program for reducing carbon dioxide emissions.[16] Policy challenges involving land-surface management of carbon dioxide are compounded by fundamental uncertainties in carbon-cycle science, including a basic understanding of the contribution of human perturbation to forests to global emissions.[17] The capture and storage of carbon dioxide will be revisited in Chapter 5. The remainder of this chapter will focus on carbon dioxide emissions that result from the combustion of coal, petroleum, and natural gas.

Figure 1.3 shows the annual increase in atmospheric concentrations of carbon dioxide from 1959 through 2008, over which reliable measurements have been taken following the International Geophysical Year in 1957. The measurements have been taken from near the summit of the Mauna Loa volcano on the Big Island of Hawaii, starting in March 1958 and continuing to the present. The increasing concentrations are analogous to the water level in the metaphorical bathtub described above.

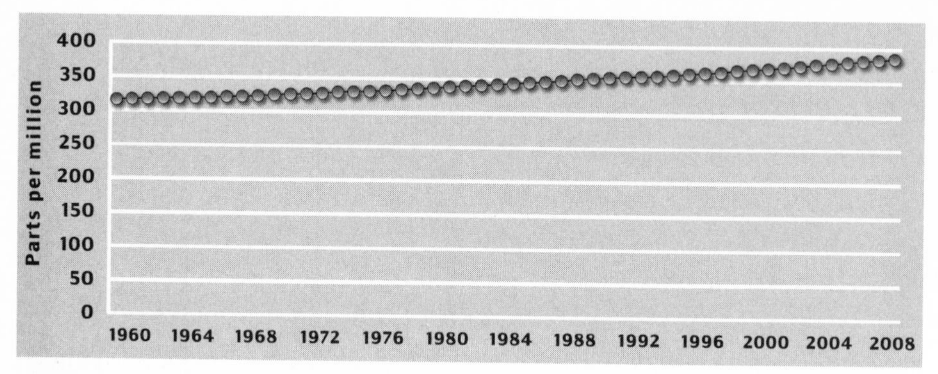

FIGURE 1.3 Atmospheric carbon dioxide concentrations, 1959–2008. Source: U.S. National Oceanic and Atmospheric Administration.

Policy makers have debated goals for stabilizing carbon dioxide–concentrations and for the amount of emissions reductions that policies should seek to target through climate policy.[18] In other words, they are debating two quantities at once: one quantity is how high the top of the bathtub actually is, that is, when the water spills over and causes damage (called a stabilization target), and the second quantity is what level they should seek to limit the water's increase and the milestones on the way to achieving that limit (called emissions-reduction targets). Here the bathtub analogy breaks down, because the impacts of carbon dioxide are not instantaneous, like the effects of a bathroom flood. Carbon dioxide actually affects the atmosphere, oceans, and ecosystems at all concentrations. Although some scientists believe that there may be "tipping points" or thresholds in the climate system where catastrophes occur, there inevitably remains much that is unknown. So unlike the bathtub, in the real world the impacts of increasing carbon dioxide are already occurring, and no one knows if or when there might be a threshold effect.[19] To make the bathtub analogy a bit more realistic you'd have to imagine water filling up a bathtub when you don't know the level of the rim or even the size of the tub. As Kerry Emanuel, an MIT climate expert who focuses on tropical cyclones, explains, "I do not think there is any 'magic number' that denotes some kind of tipping point, but if there is, we collectively have no idea what number that is."[20]

Because of the inevitable and fundamentally irreducible uncertainties about the future impacts of accumulating carbon dioxide, policy makers have sought to define a political threshold to guide policy, informed as much as possible by information from the climate science community. Politics works well with nice, round numbers, so a number that was first discussed in early climate policy discussions was 550 parts per million, simply because this was a doubling of preindustrial concentrations, and thus easily conveyed. It was also the value most frequently used by climate scientists in their early climate-modeling work, so 550 ppm was a number for which there was actual research to discuss.[21]

In more recent years the target most often mentioned in climate policy is 450 ppm, which has been advocated by the European Union as being consistent with limiting global average temperature changes to 2 degrees Celsius above preindustrial values. The 2 degree target has been severely criticized by some experts. Among them is polymath Richard Tol, a Dutch economist, who has written that the target is "supported by rather thin arguments, based on inadequate methods, sloppy reasoning, and selective citation from a very narrow set of studies."[22] Tol suggests that the 450-ppm target may simply be a negotiating position in the international climate policy process, with less basis in science than in the convenience of being a round number. Regardless, the value of 450 ppm, used interchangeably with a "2 degree target," has been at the center of international negotiations in recent years.

Others find the 450-ppm target to be far too high. For instance, author and environmental activist Bill McKibben founded an organization called 350.org to advocate for stabilizing concentrations at no higher that 350 ppm, which would imply removing carbon dioxide from the atmosphere in order to reach a level last seen in the 1980s. McKibben explains why 350 makes more sense than 450, echoing some of the same concerns raised by Tol: "Science doesn't actually know if 450 ppm and 2 degrees are the same thing, and no one knows how much change they would produce. Again, these were guesses for the point at which catastrophic damage would begin—they were more plausible, but still not based on actual experience. They also reflected guesses of what was politically possible to achieve." McKibben further explains that a 350-ppm target offers greater political traction in the debate than would a 450-ppm target: "It's the difference between a doctor telling you that you really should think about changing your diet and a doctor telling you your cholesterol is already too high and a heart attack is imminent. The second scenario is the one that gets your attention."[23] So 350 is a round number like 450, but it has the added benefit of implying urgency, as the atmospheric concentration of carbon dioxide has already exceeded that value.

Among the members of McKibben's 350.org organization is Rajendra Pachauri, chairman of the Intergovernmental Panel on Climate Change, who has stated that "what is happening, and what is likely to happen, convinces me that the world must be really ambitious and very determined at moving toward a 350 target." Also supporting a 350-ppm target is NASA's James Hansen, who has written, "If humanity wishes to preserve a planet similar to that on which civilization developed and to which life on Earth is adapted, paleoclimate evidence and ongoing climate change suggest that CO_2 will need to be reduced from its current 385 ppm to at most 350 ppm."[24] At the UN climate meeting in Copenhagen in December 2009, McKibben announced that ninety-two small, mostly poor nations had endorsed calls for a 350-ppm target.[25]

Whether a 350- or 450-ppm target makes more sense is largely a distraction to the challenges of policies focused on stabilizing concentrations of carbon dioxide in the atmosphere. Ken Caldeira, a prominent climate scientist at Stanford University, explains that the focus should be on the flow of emissions, not the ultimate concentration target: "I think that arguments over temperature targets are a distraction. We should be talking about emissions targets, and the right emission target is zero. We are going to solve the carbon-climate problem when we create an understanding that it is no longer acceptable to use the atmosphere as a waste dump."[26] Caldeira is simply explaining the logic of the bathtub model: the rise in concentrations won't stop until the amount of carbon dioxide going into the atmosphere equals the amount going out.

Reaching this goal implies virtually the same set of actions regardless of whether the target is 450 ppm, 350 ppm, or some other low level for stabilization. Myles Allen, a climate researcher at Oxford University, puts it well: "The problem is not that 350 ppm is too high or too low a threshold, but that it misses the point. The actions required over the next couple of decades to avoid dangerous climate change are the same regardless of the long-term concentration we decide to aim for."[27] Consequently, the debate over targets is a little like arguing whether we should seek to advance the average human life span to 87.5 years or 93.5 years. Surely, improving health outcomes is an important public goal, but progress occurs when we develop solutions to disease and public health issues, not when we agree on targets for human life spans. But I am getting ahead of myself, as there is still some additional science to be discussed.

The effects of carbon dioxide on the climate system are enhanced by a process called water-vapor feedback, which describes how the atmosphere holds more water when it warms. Because water vapor is itself a powerful greenhouse gas—so named by scientists to try to convey greenhouse gases' net warming effects—more water vapor in the atmosphere will enhance the warming effects of added carbon dioxide. In 2001 the Intergovernmental Panel on Climate Change wrote that water-vapor feedback was "the most consistently important feedback accounting for the large warming predicted by general circulation models" in response to accumulating carbon dioxide in the atmosphere.[28]

Water-vapor feedback only begins to hint at the enormous complexities of the global climate system. Figure 1.4 shows how some of these complexities were presented in a 1974 paper by NCAR's Will Kellogg and Steve Schneider. As understandings have developed over the years, it has become apparent that the climate is an even more complex system than Kellogg and Schneider indicated. In 2007 the IPCC referred to a much wider range of mechanisms and pathways of feedback that did not appear in the Kellogg-Schneider figure when it wrote that "the climate system is a complex, interactive system consisting of the atmosphere, land surface, snow and ice, oceans and other bodies of water, and living things."[29] In short, just about every atmospheric, oceanic, and ecological process on planet Earth is a part of the climate system. Table 1.1 provides a list of many of the factors related to human activities that are now thought to have a discernible effect on local, regional, and global climate over time periods ranging from weeks to centuries. These factors interact with each other and lead to second-order

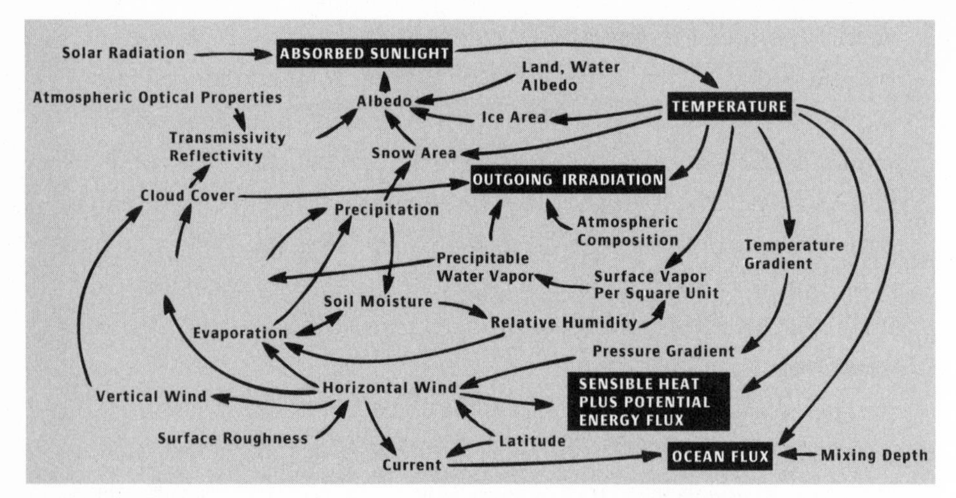

FIGURE 1.4 Climatic cause-and-effect (feedback) linkages. Source: W. Kellogg and S. Schneider, "Climate Stabilization: For Better or Worse?" *Science* 186 (1974): 1163–1172.

(and third-and so on) effects that also feed back to the system, influencing the climate system in a diverse set of ways.

Despite the wide range of human-related climate influences, scientists and policy makers have focused their attention on the climate-altering roles of greenhouse gases, and among greenhouse gases the focus has been primarily on carbon dioxide. For instance, in 2007 the IPCC reported, "Carbon dioxide is the most important anthropogenic greenhouse gas." John Holdren, science adviser to President Barack Obama, has written, "Carbon dioxide is the most important of civilization's emissions and the most difficult to reduce." Justice John Paul Stevens wrote in a 2007 U.S. Supreme Court decision in which it ruled that carbon dioxide and several other greenhouse gases can be classified as dangerous pollutants that carbon dioxide is "the most important" greenhouse gas.[30]

Carbon dioxide is important not just because of its influence on climate but also because it is expected to stay in the atmosphere much longer than other greenhouse gases.[31] Susan Solomon, who led the science report for the IPCC in 2007, and colleagues have written that "greenhouse gases such as methane [CH_4] or nitrous oxide [N_2O] are significant for climate change in the next few decades or century, but these gases do not

TABLE 1.1 Human influences on the atmosphere

Greenhouse gases

CO_2	CH_4	N_2O	CFC-11	CFC-12	CFC-113
HCFC-22	HCFC-141b	HCFC-142b	CH_3CCl_3	CCl_4	HFC-125
HFC-134a	HFC-152a	HFC-23	SF_6	CF_4 (PFC-14)	C_2F_6 (PFC-116)

Tropospheric ozone	Stratospheric ozone
Stratospheric water vapor	Tropospheric water vapor

Atmospheric aerosols

Black carbon (soot)	water-soluble inorganic species: e.g., sulfates, nitrates (reactive nitrogen), and ammonium	condensed carbon species	mineral dust from land degradation
Biomass burning	Other particulate matter: e.g., volatile organic chemicals, metals, and soil or dust particles	Airplane contrails	

Variable properties of aerosols

Irregular distribution of aerosols, which results from their short lifetime (a few days to a week);	chemical composition, especially the organic fraction;	mixing state and behavior (tendency to absorb water, density, reactivity, and acidity)
Optical properties associated with mixing and morphology (refractive index, shape, solid inclusions)		

Land-use change

Urbanization	desertification	deforestation	afforestation
Overgrazing	irrigation	dryland farming	changes to albedo

Sources: http://www.ipcc.ch/pdf/assessment-report/ar4/wg1/ar4-wg1-chapter2.pdf; http://www.nap.edu/openbook.php?record_id=11175&page=34.

persist over time in the same way as carbon dioxide." Stanford University's Ken Caldeira maintains that "if you say what's the primary gas responsible for the planetary warming, I would say it's carbon dioxide."[32] The public discussion of climate change echoes the importance that scientists and politicians have placed upon carbon dioxide. In the public debate over climate change we speak in terms of carbon footprints; no one seems to worry much about their nitrous oxide footprint.[33]

Because the policy debate and scientific emphasis among virtually all leading scientific assessments have been focused on carbon dioxide, the emphasis of the analysis in subsequent chapters will also be on carbon dioxide.[34] The result of this emphasis, of course, is that other important human influences on the climate system will be neglected. Some might see this as a feature: Steve Rayner, a close colleague of mine and professor at Oxford University, has suggested to me that international climate policy was off track from the moment that it centered on climate change broadly conceived rather than starting from a narrower focus on long-lived greenhouse gases or even more narrowly on carbon dioxide. His point notwithstanding, dealing only with carbon dioxide or even greenhouse gases cannot fully address the challenge of climate change, a subject that I'll return to later. A commonsense approach to climate policy must begin with recognition that carbon policy is not the same thing as climate policy, despite the conflation of the concepts in popular, and often in technical, discussions.

Make no mistake: carbon dioxide matters a great deal. However, a key implication of recognizing the diversity of human influences on the climate system is that even if we were to meet the challenge of stabilizing concentrations of carbon dioxide (and even other greenhouse gases) in the atmosphere at a low level, we would not have solved the larger challenge of addressing human influences on the climate system, because we have so many other influences on it. Although the next several chapters focus on carbon dioxide, this will not be the last time that we discuss other greenhouse gases or even other human influences on the climate system.

Some policy instruments have been designed to address carbon dioxide in the context of addressing a larger set of climate-influencing factors. For instance, the Kyoto Protocol has sought to reduce the emissions of a "basket" of six different greenhouse gases, while legislation proposed in the United States in 2009 was designed to reduce the emissions of seventeen different greenhouse gases. Even more complex policy proposals seek to manage emissions from forests, agriculture, and other land uses. Some have suggested bringing the regulation of aerosols (such as soot) into the climate debate. However, even as most of these policies and proposals can be mind-numbingly complex, none deal with the full complexity of the human influence of the climate system. As will be argued throughout this book, thinking that we might manage the entire human influence on climate under a single policy umbrella is fanciful at best, as the more complex a policy is, the more difficult it is to implement. A commonsense approach to climate policy will reflect a diversity of policies and instruments, reflecting the diversity of issues encompassed under the umbrella of "climate change."

One of the most obvious challenges facing efforts to manage the earth's climate is that as science progresses, new complexities are being discovered and proposed all the time. In the summer of 2009, for instance, two researchers from Cal Tech published a paper in the prestigious journal *Nature* in which they suggested that the movement of jellyfish in the oceans might exert a climatic effect comparable to winds and ocean currents.[35] This prompted at least one climate scientist to wonder if climate models that seek to describe and predict the evolution of the climate system might have to be modified to include the effects of such "biogenic mixing."[36] Because humans also have a large footprint on oceans and sea life, it could very well be that if human activities influence biogenic mixing, such as through fishing or pollution, then there would be yet another human influence on the climate system. The list presented in Table 1.1 will inevitably grow longer and more detailed as scientists continue to better understand the myriad complexities of the human role in the global climate system.

The presence of multiple human influences on the climate system has proved problematic at the science-policy interface, where much of the public discussion takes place as a Manichaean debate of skeptics and deniers versus consensus defenders and alarmists (with the choice of group labels a function of where you stand on the issue). This two-sided debate has been difficult territory to navigate for those espousing a more nuanced perspective than the cartoonish caricature of debate over climate science-as-a-proxy-for-political-debate found on blogs and in the media.

For instance, some scientists, such as my father, believe that "humans have an even greater effect on climate than is suggested by the IPCC. The human influence on climate is significant and multifaceted."[37] This view does not deny the important role of carbon dioxide in influencing climate, but it places it into a broader context in which there are multiple influences on the climate system, such as those listed in Table 1.1. My father has at times been castigated in public debate as a skeptic or even a denier, but it is hard to make sense of this label placed upon someone who has written, "It is clear, of course, that human activities have had and will continue to have a discernible influence on the climate system," and who has called for recognition that there are numerous human influences on the climate system: "Climate change as realized at the regional scale involves more than just the radiative effect of a global change in CO_2, and other greenhouse gas and aerosol concentrations." Specifically, in his research he has pointed to several other important factors that have an important influence on climatic patterns. For example, "at least two important forcings have been excluded in the IPCC. . . . These are the effect on global climate of anthropogenic land-cover change and the biological effect of anthropogenically increased concentrations of carbon dioxide."[38]

On the issue of the importance of human influences beyond carbon dioxide, my father has had some interesting fellow travelers, most notably James Hansen. In 2001 Hansen, having coauthored a paper on the effects of soot on the climate system, expressed similar concerns about the narrow focus of climate policy discussions. In a congressional hearing Hansen complained that the political emphasis on carbon dioxide was taking attention away from other important human influences on the climate system: "[The] Kyoto [Protocol] excludes consideration of air pollution"; as a result, the "IPCC basically ignores these topics and downgrades them." Hansen went so far as to allege that scientists at the IPCC and the prestigious journal *Nature* made decisions about what climate science to highlight based on their political preferences: "The only IPCC 'review' of our paper was by the IPCC leaders (as reported in the *New York Times,* for example), who saw our paper as potentially harmful to Kyoto discussions. They received the backing of organizations (such as the Union of Concerned Scientists . . .) and publications (particularly *Nature*), who had previous editorial positions favoring the Kyoto Protocol."[39]

Hansen became a darling of advocates seeking action on carbon dioxide when he testified in a court case in the United Kingdom in support of a group of individuals who had been arrested for causing damage to a coal plant, protesting its carbon dioxide–rich emissions. Hansen was just as quickly deemed "irrelevant" by many of those very same advocates when he came out in strong opposition to "cap and trade" policies to limit carbon dioxide emissions in the United States.[40]

The foregoing isn't to say that politicians are ignoring every influence on the climate other than carbon dioxide and other greenhouse gases; black carbon (soot), for example, is on their radar screen. However, in 2009 the U.S. Environmental Protection Agency in its endangerment finding of carbon dioxide declined at the same time to classify soot as a pollutant, ironically enough citing lingering scientific uncertainties as the basis for inaction.[41] Of course, once the EPA opens the door to regulation of factors other than greenhouse gases that influence the climate system, it could inevitably lead to calls for regulation of every human influence on the climate system listed in Table 1.1. That, of course, would be a regulatory nightmare, and so provides a pragmatic reason to keep the focus centered on carbon dioxide. On the international front there has been reluctance to distract attention from emphasizing carbon dioxide. In October 2009 India's environmental minister, Jairam Ramesh, commented that black carbon had no place in international negotiations on climate change. "Black carbon is another issue," he said. "I know there is now a desire to bring the black carbon issue into the mainstream. I am simply not in favor of it."[42] Although the climate system's complexity is a scientific reality, dealing with it can be a political nightmare.

Despite the reluctance of some policy makers to consider the importance of human influences on climate beyond carbon dioxide, the scientific community has increasingly emphasized the greater complexities than are recognized in a carbon dioxide–only approach. For instance, Nobel Prize–winning scientist Mario Molina and colleagues wrote in October 2009 in the journal *Proceedings of the National Academy of Sciences* that a complementary focus to carbon dioxide reductions is important: "There is growing demand among governments and commentators for fast-action mitigation to complement cuts in CO_2 emissions, including cuts in non-CO_2 climate forcing agents, which together are estimated to be as much as 40 to 50 percent of positive anthropogenic radiative forcing" (a metric used by scientists to quantify human influence on climate system).[43] In late 2009 my father along with eighteen other fellows of the American Geophysical Union published an article in which they argued:

> In addition to greenhouse gas emissions, other first-order human climate forcings are important to understanding the future behavior of Earth's climate. These forcings are spatially heterogeneous and include the effect of aerosols on clouds and associated precipitation . . . the influence of aerosol deposition (e.g., black carbon . . . and reactive nitrogen . . . and the role of changes in land use/land cover). . . . Among their effects is their role in altering atmospheric and ocean circulation features away from what they would be in the natural climate system. . . . As with CO_2, the lengths of time that they affect the climate are estimated to be on multidecadal time scales and longer.[44]

It is not surprising that there has been resistance to the calls from Molina, Hansen, Pielke Sr., and their colleagues to look at the human influence on climate more comprehensively. In the hyperpoliticized world of climate politics, any emphasis on factors beyond carbon dioxide (and other greenhouse gases) is, for some people, a distraction. They argue that things are complicated enough already, and nuance only distracts from the clear focus that politics (stressing carbon) requires to succeed. Nuanced but arguably more accurate scientific perspectives are difficult to advance in the debate, a topic that I'll revisit in Chapter 7.

Of course, some observers of the climate debate have pointed out quite correctly that for some advocates of action the issue is not really about the specific details of the human influence on the climate system, whether due to carbon dioxide or otherwise. Rather, broader notions of sustainability and how we as many billions of people live on planet Earth are the focus. Mike Hulme has written that "the idea of climate change can touch each of us as we reflect on the goals and values that matter to us."[45] Indeed, some argue that the reality of ever-increasing carbon dioxide is a symptom of a deeper set of problems, not simply a technical condition to be managed. In the days before the December 2009 climate conference in Copenhagen, Yvo de Boer, head of the United Nations Framework Convention on Climate Change (FCCC), argued that climate policy was about more than reducing emissions: "It will provide the biggest opportunity since the industrial revolution to rebalance economic activity towards a more stable and equitable path for every nation."[46] More broadly, climate change has been used as a fulcrum to gain leverage on issues as varied as biodiversity loss, deforestation, poverty, equity, population growth, consumption, global governance, and many others, all of which can—arguably—readily stand on their own merits independent of concerns about climate change.

Despite the significant attention paid to disputes and debates related to climate science, there are many aspects that are accepted by just about everyone in the discussion. Andy Revkin, formerly a reporter for the *New York Times* who has covered the climate debate for more than twenty years, explains that with respect to the climate and energy challenge there are "core ideas that are powerfully established."[47] Revkin tops his working list of these core ideas with the observation that human activities are leading to an increase in carbon dioxide concentrations in the atmosphere, and this increase will have discernible effects on the climate and oceans of the earth; most notably, it will, all other things being equal, have the effect of warming global average temperatures.

Yet even though global average temperatures are often a focal point of the debate over climate science and politics, they are not as relevant as many think. Consider that even if global average temperatures were not increasing there would still be causes for significant concern about accumulating carbon dioxide in the atmosphere. Consider the following thought experiment. Divide the world up into 1,000 boxes of equal area. Now imagine that the temperature in each of 500 of those boxes goes up by 20 degrees while the temperature in the other 500 goes down by 20 degrees. The net change in global average temperatures in this scenario is exactly zero. However, the global impacts of the changes described would be enormous. It is not because of human influences on climate at some global average scale that I worry about carbon dioxide emissions, but what happens at human and ecological scales. Global average temperature has become an important political symbol, captured best in the phrase *global warming*, but it is not the main reason carbon dioxide is of concern. Regional impacts on climate, sea level rise, and changes in ocean chemistry provide plenty of reason for me to ask whether decarbonizing the global economy makes sense.

One last point about the fixation on carbon dioxide is important to recognize: these core scientific ideas about the influence of carbon dioxide on the climate system, in addition to being almost universally accepted by partisans in the climate debate, were in fact accepted both by James Hansen in his June 1988 congressional testimony as well as by Fred Singer in his rebuttal in the *Wall Street Journal*. Despite any number of disagreements about the various scientific issues associated with climate change and where the balance of evidence on those issues lies, research since that time has underscored this core understanding. Fortunately, despite the points on which our general understanding of climate science could be, and ought to be, more nuanced, a commonsense approach to climate policy requires no more agreement on climate science than on such very basic ideas. Policy makers routinely make decisions on the economy, on military action, and on regulation with a similar (or even less well-developed) state of understanding. The heat of the climate debate can obscure the fact that there is a shared understanding of the role of carbon dioxide held by everyone in the debate.[48] A narrow focus on carbon dioxide is double-edged: it gives a sense of priority to one very important aspect of the human influence on the climate system, but it can obscure the fact that the issue of climate change involves so much more. In the conclusion to this book I'll present suggestions on how we might reconcile the narrow focus with a much needed broader perspective.

Learning the Right Lessons from the Cases of Ozone Depletion and Acid Rain

Just as policy making has converged on carbon dioxide as the key scientific aspect of climate change, so too has policy debate converged on the lessons of ozone depletion and acid rain as suggesting a policy blueprint for handling climate change. As we have seen with the science, the policy context is not so simple, either.

During the summer of 1988 the scientific issue that I paid the most attention to was not climate change but ozone depletion. The scientists for whom I worked at NCAR were involved with a major research project on ozone depletion over Antarctica. Scientists involved in ozone research at the time included Ralph Cicerone, then at NCAR and today the head of the U.S. National Academy of Sciences, and Susan Solomon, who at the time was at the National Oceanic and Atmospheric Administration in Boulder and later served as the cochair of the science report of the Fourth Assessment of the Intergovernmental Panel on Climate Change (and is now a colleague of mine at a cooperative research institute at the University of Colorado and NOAA).

The global response to ozone depletion is often invoked as a direct policy precedent for dealing with increasing concentrations of carbon dioxide in the atmosphere. Conventional wisdom holds that the science was made certain and then an international protocol was negotiated, leading to the invention of technological substitutes for chlorofluorocarbons.[49] I first had a hint that the conventional wisdom on the ozone case might be flawed when while working on my dissertation in the early 1990s I explained it to my friend and mentor Rad Byerly, for whom I had worked at the House Science Committee. Rad quickly disabused me of these myths by explaining that he was directly involved in the drafting of ozone legislation in the 1970s, and the role of science was not at all as academics had encapsulated the experience. This taught me another lesson: always check elegant academic theories against data from the real world. The ozone case does provide important lessons for understanding climate policy, but they are not the ones typically drawn.[50]

In 1974 Mario Molina and Sherwood Rowland published a seminal paper in *Nature* in which they argued that chlorofluorocarbons posed a threat to Earth's ozone layer. Ironically, CFCs were long considered to be a useful industrial chemical for a wide range of applications, including refrigeration, because of their inert properties. Molina and Rowland's work suggested that these chemicals were not as inert as previously thought.

Following the publication of the *Nature* paper, the U.S. Congress went to work almost immediately, initiating hearings before the end of the year. The White House, under President Gerald Ford, set up the Inadvertent Modification of the Stratosphere (IMOS) Task Force, which concluded that "fluorocarbon releases to the atmosphere are a legitimate cause for concern" and recommended that "the federal regulatory agencies initiate rulemaking procedures for implementing regulations to restrict fluorocarbon use."[51] Congress proceeded incrementally, first dealing with nonessential uses for CFCs, that is, those for which there were readily available technological substitutes, and putting off until later the more difficult issue of essential uses, those for which no substitutes were available. Policy makers had decided that action on the problem of ozone depletion could not wait until scientists reached consensus about the nature of the problem, its causes, and its future impacts. Decisions would have to be made in the face of uncertainties and ignorance—where even uncertainties were unknown.

As Congress made decisions about the chemicals implicated in ozone depletion in the late 1970s and early 1980s, the science of ozone depletion actually became more uncertain, as scientists began to understand the many complexities of the issue. In 1982 the National Academy of Sciences released a report suggesting that the threat of ozone depletion was perhaps less than previously thought, which was seized upon by some in Congress to argue against regulation of CFCs. There were plenty of people who were skeptical about the magnitude of the ozone problem who were buoyed by fundamental uncertainties in the science. However, the focus on implementing "no-regrets" policies—those that made sense anyway, regardless of how scientific uncertainties broke in the future—kept the attention off science and on policy options. Such an approach contributed to the invention of substitutes for CFCs, making political action all the more easier, as the justifications for action hinged less on scientific certainties and more on economic benefits.

According to the official UN history of the ozone issue, there were exceedingly few news stories on ozone depletion in the United States, China, the United Kingdom, and Soviet Union from 1977 to 1985, when much of the policy framework for the issue was developed. The *New York Times* had about twenty stories in 1982, and in no other year were there that many stories (cumulatively) in ten different leading newspapers during that period.[52] This was also a time of intense (and legitimate) scientific debate. In fact, many people believed that after the aerosol spray-can ban in the late 1970s that the problem had been solved. As policy makers dealt with the ozone issue, it had nowhere near the salience and strength of opinion that we now see among the public on climate change (see Chapter 2 for a discussion).

Action on ozone proceeded incrementally, with many decisions made step by step, first in the United States and then internationally. There was no "threshold for action" in public opinion or other metrics of "political will" that we see so often called for in the context of climate change. Action took place based on what the political dynamics would allow. Science, of course, played a very important role in placing ozone depletion on the decision-making agenda and then again in providing information useful for the fine-tuning of the international protocol once it had been widely accepted. Between these times it was effective politics and a healthy policy process that made progress possible in phasing out the production of CFCs, not science, and certainly not a unified vision of the science.

In short, on ozone depletion policy makers used science as an indicator of a possible problem and then very much followed a no-regrets strategy, taking on what was relatively easier first and leaving the more politically difficult challenges for later. In this way they reduced the scope of the problem, making debates about science less relevant and also reducing the intensity of the political obstacles to action. Policy responses to climate change have seen the opposite strategy, with the most difficult challenge and largest framing (regulating global energy) at the center of the debate. Consensus science really did not play a major role in ozone policy until after the Montreal Protocol was established in 1987, when the issue was mature and fine-tuning was possible in the details of the policy responses, largely outside of public view. The so-called ozone hole was discovered in 1985, after the U.S. domestic response was in place and after the international approach had been codified in the Vienna Convention.

Experience with policy responses to acid rain shares many characteristics with that of ozone depletion. Political action in the United States, expressed in the Clean Air Act amendments of 1990, took place before the nation's major scientific assessment had even been completed. This is not to argue that science was irrelevant, but rather it was the political dynamics that enabled the issue to reach resolution, not a unified vision of the science. In this case, the certainty in the costs of technology needed to clean up smokestack effluent and of substituting low-sulfur coal from the western United States facilitated reaching a political agreement.[53] If reducing carbon emissions was as simple as the switch from CFCs to CFC replacements, or installing scrubbers on smokestacks and sourcing Wyoming coal, we would not be discussing the issue, as it would have been mostly solved already.

The political discussion of climate change has learned the wrong lessons from the responses to ozone depletion and acid rain. Progress on climate change mitigation might be more effective if we learned the right ones. Although some observers accurately point out that the problems of ozone depletion and acid rain, because of their relative simplicity, do not provide much guidance about how to respond to climate change, the lessons that we should take from the ozone and acid rain cases are that certainty in science is unnecessary for action and that policy that proceeds incrementally can work to reduce the scope of a problem, making the politics easier while addressing the problem bit by bit. These are not the lessons that have been applied in climate policy.

Prediction Means Something Different for the Weather, Earthquakes, and Black Swans

One thing that I suppose is inevitable to learn when growing up with a father who is a meteorologist is the limits of prediction. I cannot remember a time in my life when I did not know what the weather was expected to be many days in the future. When I was a child, long before the Internet, a stop at the Atmospheric Sciences Department was a mandatory part of Pielke family outings in order for my father to take a look at the latest weather maps hot off the printer. But you don't have to be the son of a weather forecaster to know that such forecasts are highly uncertain. It is a strength of science that predictions are made and then evaluated, because such a process can help to inform where your knowledge may be wrong or incomplete. At the same time, knowledge may be exceedingly strong, yet predictions perform poorly.

But given that weather forecasts are not always right, our experience with them, it turns out in an ironic sort of way, actually enhances their value in decision making. Decision makers, including most of us as individuals, have enough experience with weather forecasts to be able to reliably characterize their uncertainty and make decisions in the context of that uncertainty. In the United States the National Weather Service issues millions of forecasts every year. This provides an extremely valuable body of experience for calibrating forecasts in the context of decisions that depend upon them. The remarkable reduction in loss of life from weather events over the past century is due in part to improved predictive capabilities, but just as important has been our ability to use predictions effectively despite their uncertainties.[54]

In contrast, consider earthquake predictions, which scientists are perfectly capable of issuing. However, in the 1970s it was decided that U.S. policy responses to earthquakes would focus not on prediction but instead on a no-regrets approach concentrated on increasing resilience of cities in places where earthquakes are known to occur. Thus, independent of the ability to predict earthquakes, cities in the United States typically do very well when earthquakes occur due to plans put in place that will succeed no matter when an earthquake strikes. (Of course, building resilience in communities is a constant challenge, and every earthquake event shows where we could have done better.) Another reason policy makers don't rely on earthquake prediction is that the costs of a false alarm can be very high, both in terms of societal response and in terms of the credibility of the scientific community.

Unlike weather forecasts, earthquake predictions could never result in enough accumulated experience to allow people to form valid judgments about their inherent uncertainty and reliability and thus learn how to behave accordingly. We all know how to respond to a forecast of a 60 percent chance of rain; knowing what to do with a prediction of a 60 percent chance of an earthquake is much less clear.

The examples cited above for weather and earthquake prediction come from a project that I was involved in during the late 1990s focused on prediction in the earth sciences and its role in decision making.[55] We looked at ten different cases to distill a set of heuristics to help better understand when decisions should and should not rely on specific predictions about how events will turn out as the basis for decision making. Here are five criteria for when decision makers should rely on predictions:

1. Predictive skill is known.
2. Decision makers have experience with understanding and using predictions.
3. The time period between the prediction and the event is short.
4. There are limited alternatives to prediction.
5. The outcomes of various courses of action are understood in terms of well-constrained uncertainties.

It turns out that long-term climate predictions (or projections) fail with respect to each of these criteria, whether they are expressed as certain forecasts or as probabilistic scenarios. It would take 3,000 years to obtain as much knowledge of the skill of 100-year climate forecasts as we obtain every month with respect to individual daily weather forecasts. The scientific community has indicated that a consequence of human influences on the climate system could be profound and costly, or perhaps more benign. Steve Schneider has explained a "lingering frustration" that "uncertainties so infuse the issue of climate change that it is still impossible to rule out either mild or catastrophic outcomes, let alone provide confident probabilities for all the claims and counterclaims made about environmental problems."[56]

The climate debate is such that one can find scientists who are adamant that they can predict the long-term climate future with accuracy and precision, and this group of scientists includes those making predictions that are incompatible with each other. At the same time, there are also scientists who think that climate prediction with precision and accuracy is simply beyond our abilities.

Such a situation presents us with competing certainties, irreducible uncertainties, and areas of complete ignorance. Such a diversity of views leads to a situation that science policy expert Dan Sarewitz has characterized as an "excess of objectivity" in which there is a sufficient distribution of scientific perspectives to allow political advocates to pick and choose based on political or ideological convenience.[57] One response to such a circumstance has historically been to invest more money in climate research in hopes that scientists might converge on a single view of the future, to reduce uncertainties. However, one perhaps counterintuitive consequence of increasing research, at least in the short term, is that bringing in more perspectives, more methods, and more scientists can easily lead to more uncertainties.[58] This situation would seem to call out loudly for robust, no-regrets-type decision making in the face of irreducible uncertainties and ignorance; in fact, as I will discuss in greater depth in Chapter 7, demands for certainty in climate science have contributed not to finding a solution but to the problematic politicization of climate science.

Nicolas Nassem Taleb, author of *The Black Swan*, has helpfully explained his views of climate prediction as follows: "Climate experts, like banking risk managers, have failed us in the past in foreseeing long term damages and I cannot accept certainty in a certain class of nonlinear models. . . . [O]ne does not need rationalization with the use of complicated models (by fallible experts) to the edict: 'do not disturb a complex system' since we do not know the consequences of our actions owing to complicated causal webs."[59] Much of my father's work echoes similar themes. Prediction is difficult, especially, as someone once said, about the future. John Beddington, science adviser to the UK government, said something similar in early 2010: "It's unchallengeable that CO_2 traps heat and warms the Earth and that burning fossil fuels shoves billions of tonnes of CO_2 into the atmosphere. But where you can get challenges is on the speed of change. When you get into large-scale climate modeling there are quite substantial uncertainties. On the rate of change and the local effects, there are uncertainties both in terms of empirical evidence and the climate models themselves."[60] Even with uncertainties about the future, there is ample evidence, broadly accepted, that humans are influencing the global earth system. Such influences carry with them a risk of undesirable outcomes.

Although there is much interesting science still to be done in the area of climate science, from the standpoint of policy making there is little point in arguing about alternative scientific predictions of the distant future not only because such debates are colored by more than a small amount of political preferences but also because such alternative visions simply cannot be resolved empirically on the timescale required of decision making. At the same time it is necessary to recognize that decisions are in their own way a form of prediction. We take one fork in the road rather than another because we expect that it will take us to a desired destination. Climate science does not tell us what to do or even that we have to act. It has very helpfully let us know that there are decisions that we might consider making. As subsequent chapters will argue, accelerating decarbonization of the global economy and improving adaptation to climate make good sense quite independent of long-term predictions of the climate future.

Guidelines for a Commonsense Approach to Climate Policy

The brief overview in this chapter summarizes the key aspects of climate science that I see as a necessary underpinning for a commonsense climate policy. Climate science is a rich and diverse field of study that goes far beyond the cursory discussion in this chapter. For those interested, it can be very rewarding to learn more about the science. However, decisions about what actions to take on what timescales cannot be resolved through appeals to science.

Here is a summary of the key points to take from this chapter:

Increasing Carbon Dioxide Influences the Climate System, Perhaps Dramatically and Irreversibly

That human activities have led to changes in the earth system is broadly accepted. So too is the possibility that such changes could lead to undesirable outcomes in the future. For those wanting to know more—much more—about aspects of climate science, the report of Working Group I of the IPCC is an excellent place to start further investigations, even as aspects of that report continue to be contested.

The Climate System Is Subject to Multiple Human Influences

Carbon dioxide will be the focus of the next several chapters, but it is not the only important human influence. The climate system is complex and is still not fully characterized. Even so, many scientists and policy makers have concluded that dealing with carbon dioxide should be a top policy priority; hence, that is the focus of the subsequent chapters.

Our Ability to See the Future Is Limited

There are debates about how the future will play out that simply cannot be resolved on the timescales of decision making. Efforts to gain clarity about the future may in fact have the paradoxical consequence of making that future even cloudier. Decisions about climate change will occur in the context of contestation, uncertainties, and ignorance.

Certainty Is Not Forthcoming

As decisions are made about decarbonizing economies and improving adaptation to climate in the coming years, certainties about the long-term climate future are not forthcoming. UK science adviser John Beddington explains, "There is a fundamental uncertainty about climate change prediction that can't be changed." As Andy Revkin summarizes his years of covering the climate debate: "What the debate comes down to is not whether changes are coming but when they'll occur—and how severe they'll be. There is serious scientific disagreement about such vital questions as how fast and far temperatures, seas, and storm strength could rise."[61] Such disagreements will persist for the foreseeable future.

Uncertainties and ignorance are a reality to be lived with and managed. They are not going away.

Stabilizing Atmospheric Concentrations of Carbon Dioxide Does Not Stop Climate Change

Carbon policy is not a comprehensive climate policy. It is possible that the world could successfully address accumulating concentrations of carbon dioxide in the atmosphere and still have to deal with a significant issue of human influences on the climate system. For this reason, among others, Mike Hulme has written that climate change is a problem to be managed, not solved. Our debates about climate change would benefit by distinguishing carbon policies from greenhouse gas policies and broader conceptions of climate policy.

Science is a powerful tool with which to understand the world, but it is one that requires humility in its application. Political scientist Harold Lasswell once argued that while we are all to some degree blind about the world, science and systematic study can to some degree reduce our degree of impairment. Policy making is about shaping the future, about why taking one fork in the road will take us to a better destination than another. The remainder of this book seeks to characterize different forks in the road, how we might collectively go about deciding which route to take into the future, and ultimately why one path looks to make far more sense than the others.

We leave this chapter behind knowing that achieving consensus on many aspects of climate science is unnecessary for action to take place. The scientific consensus that held in 1988 between S. Fred Singer and James Hansen was enough to form the basis for action. The more difficult question, of course, is, what action?

CHAPTER 2

What We Know for Sure, but Just Ain't So

THE POLICY ACTIONS that are considered and ultimately are adopted in climate change are shaped profoundly by some very important and often unquestioned assumptions. Some aspects of such conventional wisdom are, to be absolutely direct, just plain wrong. It should be no surprise that if key assumptions that underlie policies on climate change are wrong, then it would go a long way toward explaining why it is that climate policies have been mired in gridlock, with no real prospects for getting unstuck.

One of Al Gore's favorite sayings comes from Mark Twain: "It ain't what you don't know that gets you into trouble. It's what you know for sure that just ain't so." He uses this saying to characterize some "inconvenient truths" about the science of climate change, which Gore argues that his political opponents have refused to accept, leading to the biggest inconvenient truth of all: we are going to have to change course if we are to deal with the challenge of climate change. But views on science aren't the only inconvenient truths in the climate debate. If Gore's political opponents are guilty of misunderstanding climate science, Gore and many others focused on motivating action on climate change are equally guilty of misunderstanding key elements of the policy context.

This chapter focuses on three important assumptions about decarbonizing the global economy that are firmly entrenched as conventional wisdom but, as Mark Twain would say, just ain't so:[1]

1. We lack political will.
2. We must trade off the economy for the environment.
3. We have all the technology we need.

This chapter argues that whatever else we believe about climate policies, being wrong on just these three assumptions has been sufficient to lead to continued policy failure. Until we correct what "just ain't so" in key assumptions, policies focused on decarbonizing the economy will have virtually no hope for success.

Do We Lack Political Will?

It is a common perception that the failure of climate policies to date stems from a lack of political will. For instance, in July 2009 Al Gore repeated some of his favorite lines during a speech on climate change at Oxford University: "We have everything we need except political will, but political will is a renewable resource. . . . The level of awareness and concern among populations has not crossed the threshold where political leaders feel that they must change. The only way politicians will act is if awareness raises to a level to make them feel that it's a necessity."[2]

If the concept of political will is to have any practical meaning, it cannot simply be based on an argument that because certain climate policies have thus far failed to be implemented there must be a lack of political will, as that would mean assuming rather than proving a causal linkage. So here I'll define political will much as Al Gore has in the quote above to be represented by the public's awareness and concern about a problem and the public's expressed desire for action on that problem, as measured by formal polls of opinion. Fortunately, there is a large literature on the role of public opinion in policy making on environmental issues and beyond. This literature tells us that the relationship between public opinion and policy action is complicated, to say the least. Such complexity provides an important clue in making sense of the data on public opinion that follow.

The first question to ask is: is it in fact the case that the public has little concern about climate change or is not supportive of action? The answer is an unequivocal no, for both the United States and beyond. In fact, few issues of global concern have a track record of sustained recognition of a problem and support for action as does the climate issue.

Consistently, since at least the mid- to late 1990s, opinion polls show that about 70 percent or more of the U.S. public, with occasional excursions above and below this range, believe that humans have an influence on the climate system.[3] This view holds strong even into 2010 following a cold winter and public concerns about climate science.[4] In 2008 Gallup found "a sizable increase over the past decade in the percentage of Americans who agree that 'most scientists believe that global warming is occurring,' from 48 percent in 1997 to 65 percent this year. Republicans, independents, and Democrats have all become more likely to see most scientists as believing in global warming."[5] However, by early 2010 this number had taken a dip to 52 percent, which Gallup attributed to concerns over climate science, a particularly cold and snowy winter in the United States, and the increasingly partisan nature of the issue.[6] In the United States there have been important partisan differences in opinion, which fall along predictable lines.

In late 2009 public faith in scientists took a hit when a large number of e-mails and other materials were released or stolen from the Climatic Research Unit at the University of East Anglia in the United Kingdom. The e-mails showed scientists talking to each other about how to avoid disclosing information about their research and their desire to influence scientific publication processes in their favor and against those they disagreed with, among other dubious conversations. Soon thereafter, allegations were made about errors and conflicts of interest involving the IPCC. The fallout from these events was an immediate loss of credibility among the public, with one poll taken a week after the e-mail release showing 59 percent of the American public believing that it was at least likely that climate scientists had falsified their research in support of their views on global warming.[7] The portion of the public who believe that scientists have overstated the case for action has grown dramatically in recent years, across party lines (see Chapter 8).

Somewhat remarkably, even with partisan differences and a diminished view of the credibility of climate scientists, public opinion polls on

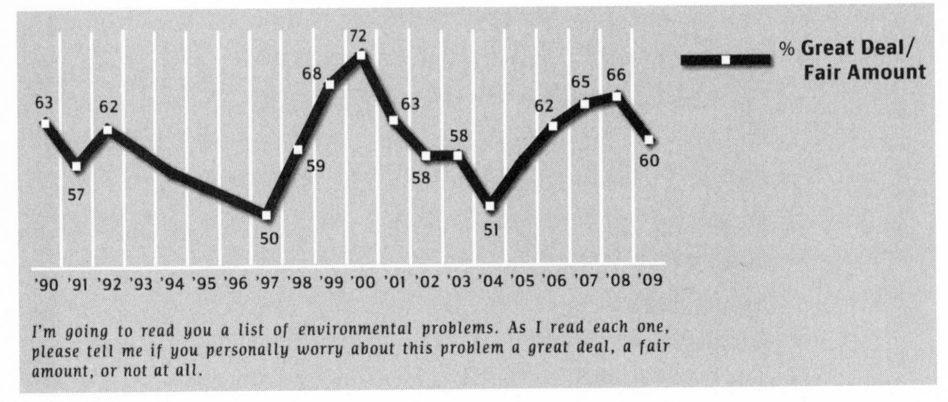

FIGURE 2.1 Public opinion on the "greenhouse effect" or global warming, showing percentage of respondents expressing a "great deal" or "fair amount" of worry about the issue. Source: Gallup.

climate change consistently show high public concern about climate change and a strong desire to take action, even as support has increased and decreased at various points in time. Figure 2.1 shows data from a 2009 Gallup poll for the United States,[8] and includes the results of polls taken over the previous twenty years, asking the public how much they worried about climate change.[9] There is a remarkable consistency over time in the public's response, with strong majorities expressing concern: 63 percent of respondents expressed concern in 1989 and 60 percent in 2009, varying by no more than about 10 percent higher or lower over that period. The Gallup analysis concludes, "Although there have been fluctuations on this measure of worry over the years, the percentage of Americans who worry a great deal about global warming is no higher now than it was nineteen years ago."[10] Remarkably, even with enormous changes in the political context of climate change in the several decades since the issue first came to occupy the public stage, the number of people expressing concern about the issue has remained fairly consistent, even with its ups and downs, across a range of polls and questions.[11]

Similarly, the number of people expressing support for action has remained fairly constant over the time period for which opinions have been polled. Figure 2.2 shows the proportion of poll respondents who were asked in a poll focused on global warming if they supported action concerning the environment, with very strong majorities supporting some or

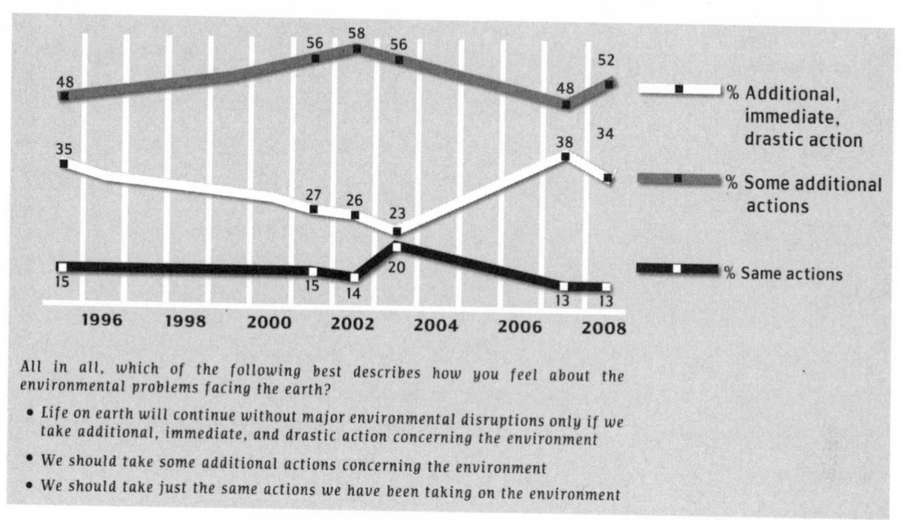

FIGURE 2.2 Feelings about the environmental problems facing the Earth. Source: Gallup.

drastic action: 83 percent in 1996 and 86 percent in 2008. The proportion calling for some action versus drastic action dipped during the early years of the presidency of George W. Bush following 9/11 and rebounded to levels identical to 1996 by 2008. The support for action has deep roots: as early as the late 1980s there was a public preference for immediate action rather than more research on the matter alone.[12] After dip in the early 1990s, which communication experts Matt Nisbet and Theresa Myers attribute to public calls for more research among political leaders and concerns about economic costs, support for action rebounded strongly and has remained strong. Consider that President George W. Bush terminated U.S. participation in the Kyoto Protocol despite overwhelming disapproval of that decision among the American public.[13] While we should expect public opinion to have its up and downs in coming years, it would only be surprising if support for action fell to historically unprecedented levels and stayed there, which seems unlikely.

The difference between the expressed public support of the Kyoto Protocol and the actions of the Bush administration hints at the complex relationship between public opinion and policy actions. Policy makers don't always do what the public would seem to prefer, and, conversely, the public doesn't always believe what policy makers would like them to. Mike Hulme writes, "Surveys of public opinion on both sides of the Atlantic about man-made climate change continue to tell us something politicians know only too well: The citizens they rule over have minds of their own."[14]

The broader international picture of public opinion is, of course, more complex, with a wide variation in public awareness of climate change. For instance, in India, where hundreds of millions of people have no regular access to electricity, only 35 percent of people were aware of climate change during 2007–2008, whereas 99 percent were aware of the issue in Japan.[15] A poll taken in late 2009 of views across fifteen countries, sponsored by the World Bank, found that in each country the public believed climate change to be a serious problem, as shown in Figure 2.3.[16] Across all fifteen countries the public expressed a strong desire for action, despite varying degrees of belief in a scientific consensus on climate change. Opinion in the United States, compared to that in other countries, arguably favors action on climate change least; nevertheless, support is overwhelmingly strong. There is no indication that public opinion is a limiting factor in the adoption of climate policies, in the abstract, anywhere in the world. In fact, in many places appealing to the public's interest in climate has proven to be a political asset in election campaigns, as demonstrated in recent years in the United States, Japan, and Australia, among other countries.

Another way to look at the importance of public support for action is to compare support for action on climate change with support for other major policy decisions. For example, in the months before the U.S. invasion of Iraq, about 60 percent of Americans favored an invasion. Conversely, before the Bush administration's federal bailout package was passed by Congress, only 28 percent of the public supported the plan.[17] Despite vast differences in opinion, both policies were enacted. "Action" on major policies takes place under a wide range of public opinion.

This conclusion is backed up more generally through systematic analyses of public opinion and policy action. For instance, a political science analysis looked at thirty-six policies for which opinion data were available and found that Congress acted in the direction of public support only in 50 percent of the cases, with public opinion having a much

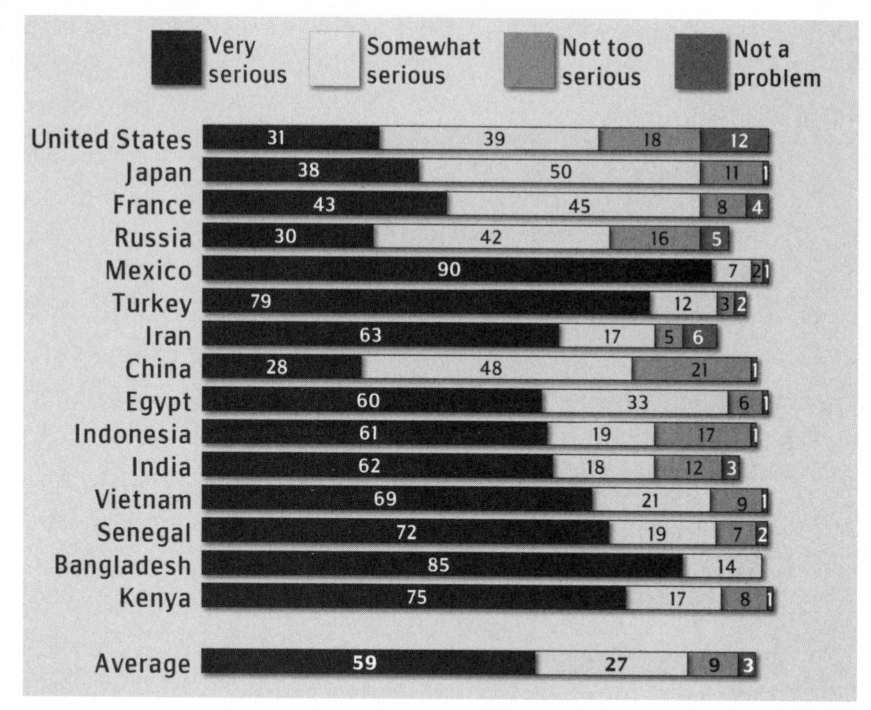

FIGURE 2.3 Is climate change a problem? Source: World Bank.

stronger influence on Congress in the direction of that opinion when it opposed an action rather than when it supported an action.[18] These data are suggestive that public opinion on climate change is well within that zone of acceptability sufficient to allow action to occur. Put another way, public opinion on climate is not a fundamental limiting factor holding back action. Political will is not lacking.

These findings are consistent with what has been observed in environmental policies more generally. In 1989 Riley Dunlap looked back over the previous several decades at the influence of the environmental movement on the adoption of environmental policies and found that a remarkable "revolution" had occurred: "The environmental movement has been enormously successful in gaining the approval and participation of the American public, probably more so than any other contemporary social movement."[19]

Dunlap's historical review of public opinion on environmental issues, when considered in the context of recent opinion polls on climate change, indicates something rather remarkable: all of the major U.S. environmental legislation of the 1970s and 1980s was implemented with a degree of public support about the same as or, in most cases, less than that which has existed on climate change for more than a decade. For instance, in a poll taken in December 1987 and January 1988, the issue of ozone depletion ranked fourteenth on a list of twenty-eight environmental problems; at the time fewer than 50 percent of Americans expressed "serious concern" about the issue, falling behind concerns about issues such as farm runoff and contaminated tap water.[20] Even so, the United States had signed on to the Montreal Protocol several months prior and ratified the treaty a few months later. The fact that public opinion on ozone depletion was not particularly intense as compared to other environmental issues provides compelling evidence that an issue does not have to be a top public priority for significant action to occur. Although the data on public opinion of the ozone issue are not comprehensive, they are strongly suggestive that policy action occurred in the context of a public that was no more concerned about ozone depletion in the late 1980s than has expressed concern about climate change for at least the past decade.

Despite the fact that public opinion on climate change has been strong in both an absolute and a relative sense, it is often and accurately pointed out that the public has other priorities that take precedence. Nisbet and Myers, comparing public concern about climate change with concern about a number of other environmental issues, found that "global warming remains at the lower end of worries."[21] And in the context of broader societal issues, climate change consistently ranks lower than issues such as the economy, education, crime, terrorism, jobs, immigration, and others. In January 2010 the Pew Center for the People and the Press found that global warming ranked twenty-first out of twenty-one issues that they asked respondents to rank in importance.[22] Global warming was also ranked as the bottom priority in a list of twenty topics in January 2009.[23] The top priority in the poll both years? The economy.[24]

An important lesson to take from polls of the public is that support for action is strong but not intense. In November 2009 Ted Nordhaus and Michael Shellenberger of the Breakthrough Institute (TBI) asserted, "What is arguably most remarkable about U.S. public opinion on global warming has been both its stability and its inelasticity in response to new developments, greater scientific understanding of the problem, and greater attention from both the media and politicians. Public opinion about global warming has remained largely unchanged through periods of intensive media attention and periods of neglect, good economic times and bad, the relatively activist Clinton years and the skeptical Bush years."[25] They maintain that the issue of climate change is tailor-made to attract public support, but at a very low intensity. The data would seem to bear out this argument. Efforts to increase intensity, whether by hyping the science or seeking to scare people with apocalyptic visions of catastrophe, are more likely to turn people off than to motivate them to become politically active.

The fact is that the political will for action on climate change is and has been strong enough for action to occur. The question left unaddressed, of course, is: strong enough for what action? The challenge facing climate policy is to design policies that are consonant with public opinion, and are effective, rather than to try to shape public opinion around particular policies. Thus far, advocates of climate policy have failed in policy design, given that there has been strong support for action, albeit at a low intensity. Despite the evidence for robust public support for action, the policy gridlock that characterizes climate policy is often represented as a failure of the public, politicians, or even democracy itself. David Victor, of Stanford University, laments the lack of progress on climate policy in such terms: "One has to wonder whether this is not a failure of governments but rather a failure of people."[26]

This review of public opinion on climate change, focused mainly on the United States, leads to the following conclusions that seem to hold generally worldwide as well:

1. A majority of the public has for many years accepted that humans influence the climate system and finds that this effect is problematic.[27]
2. A majority of the public in the United States and worldwide has strongly supported action on climate change for at least a decade,

where "strong" means similar to, or stronger, than public support for action on other major decisions such as ozone depletion or military action.[28]

3. Public support for action in the United States has consistently been an absolute majority even during the presidency of George W. Bush, when most people opposed his administration's decision to pull out of the Kyoto process.
4. Despite concern about climate change as a problem and support for action, climate change does not rank high as a public priority in the context of the full spectrum of policy issues.

Given these conclusions, there would seem to be exceedingly little additional value in seeking to convince people that climate change is a threat or that action is warranted if the goal is to surpass some perceived threshold for action. As Nordhaus and Shellenberger wrote in late 2009, "Majorities of Americans have, at least in principle, consistently supported government action to do something about global warming even if they were not entirely sold that the science was settled, suggesting that public understanding and acceptance of climate science may not be a precondition for supporting action to reduce greenhouse gas emissions."[29] If there is a threshold of understanding among the public that needed to be surpassed to make action possible, that threshold has clearly been passed. Nevertheless, strong and consistent public support has not stopped many advocates for action from engaging in a vicious battle with those who express minority views about aspects of climate science.

A commonsense approach to climate policy must recognize that public opinion is not determinative of a policy outcome. The fact that public opinion on climate change has been squarely in the zone where action is possible should suggest that it is not a lack of political will that is holding back action. Further, efforts to intensify public opinion could indeed have the opposite effect if they are perceived to be misrepresenting the scientific and policy arguments for action. In fact, as I will show in Chapter 7, this is exactly what has happened.

To paraphrase the American pragmatist Walter Lippmann, the goal of politics is not to get everyone to think alike but, rather, to get people who think differently to act alike.[30] This wisdom has been ignored in a climate debate in which those advocating action too often focus on trying to get everyone to think alike, forgetting that it is how people act, not what they think, that in the end matters most.

Must We Trade the Economy for the Environment?

The belief that economic growth and action on climate change are incompatible has in some ways begun to break down. But the idea that the environment and the economy are trade-offs has deep roots. In a wide range of settings there continues to be the belief that dealing with climate change will necessarily mean that economic sacrifices have to be made. It is, of course, possible to design policies that force trade-offs between the economy and the environment. But must it necessarily be so? According to many in the climate debate, the answer is yes. Here I suggest that not only can the answer be no, but it must be the case if effective action to decarbonize the global economy is actually to occur.

That the economy and the environment are necessarily trade-offs is built into the IPCC's four families of scenarios used as the basis for its projections of future emissions.[31] The IPCC distinguishes between two sorts of trade-offs. The first is between more global and more regional integration. The second explicit trade-off is between the economy and the environment, strongly implying that these valued outcomes are incompatible with each other. These scenarios contrast two worlds: one moving toward the ideals associated with "green" political parties and "sustainability," the other focused on economic growth and technology. Not surprisingly, the ideas behind such thinking mirror long-standing political differences between the United States and Europe that have found their way into scenario development.[32] The idea that economy and environment must be trade-offs clearly involves considerations that go well beyond the climate debate.

Advocates routinely suggest that action on climate change necessarily means sacrifice. For instance, in late 2009 Kevin Anderson of the Tyndall Center for Climate Change Research in the United Kingdom argued that a "planned recession" would be necessary in the United

Kingdom in order to reduce emissions in response to the threat of climate change. In practice this would mean that "the building of new airports, petrol cars and dirty coal-fired power stations will have to be halted in the UK until new technology provides an alternative to burning fossil fuels."[33] Similarly, in a comment with more symbolic than substantive importance, Rajendra Pachauri, head of the IPCC, argued that restaurants should no longer serve ice water, as an illustration of how we need to change our lifestyles.[34] Such calls for sacrifice are a fixture in debates over responding to climate change. However, if there is an iron law of climate policy, it is that when policies focused on economic growth confront policies focused on emissions reductions, it is economic growth that will win out every time.

The iron law of climate policy reflects a powerful ideological perspective that is broadly shared. Society could have evolved around a different ideological commitment, and may yet in the future. Nevertheless, as a deeply ingrained ideological commitment, the iron law will never be easy to displace. For instance, in early 2009 Steven Chu, when nominated by President Obama to serve as secretary of energy, was asked in his confirmation hearing about earlier comments he had made about the need to increase the costs of gasoline, arguing that prices needed to be boosted in the United States to levels comparable to those in Europe, which are two, three, even four times higher.[35] At his confirmation hearing Chu quickly stepped back from his comments: "What the American family does not want is to pay an increasing fraction of their budget, their precious dollars, for energy costs, both in transportation and keeping their homes warm and lit."[36] There was no sign of his earlier advocacy of large increases in the costs of gasoline; instead, he advocated that they not increase during times of economic difficulty: "As secretary of energy, I think especially now in today's economic climate, it would be completely unwise to want to increase the price of gasoline."[37] Left unaddressed, of course, is when, exactly, it would be a good time to argue for higher-priced energy. For many if not most politicians, the answer would be never.

Secretary Chu's change of view was the result of coming face-to-face with political realities. That change of heart notwithstanding, the idea that economic growth is a necessary trade-off with efforts to reduce emissions persists, despite compelling arguments to the contrary. Nordhaus and Shellenberger, for example, argued in their book *Breakthrough* that "the satisfaction of the material needs of food and water and shelter is not an obstacle to, but rather the precondition for, the modern appreciation of the nonhuman world." This view is not, however, universally shared. For instance, in 2005 *New York Times* reporter Andy Revkin discussed on his blog pending World Bank support for the building of new coal-fired power plants in India, noting that where one stands on the desirability of these plants depends upon where one comes from: "Is all of this bad? If you're one of many climate scientists foreseeing calamity, yes. If you're a village kid in rural India looking for a light to read by, no."[38]

Noted writer and environmentalist Bill McKibben commented on Revkin's blog that one reason that he was starting a movement to limit atmospheric carbon dioxide concentrations to 350 ppm (as discussed in Chapter 1) was to call attention to a broader calculus that might imply a different answer to the question than that given by Revkin: "What if you're an Indian kid looking for a light to read by—and also living near the rising ocean, or vulnerable to the range expansion of dengue-bearing mosquitoes, or dependent on suddenly-in-question monsoonal rains[?]"[39]

Writing at the Breakthrough Institute blog, Siddhartha Shome, an engineer and senior fellow at the Breakthrough Institute, took strong exception to the implication of McKibben's comment. Shome explained that even if the most dire predictions about global warming come true, some of the poorest people in the world may still be better off tomorrow if they are able to enjoy some of the fruits of development, such as education, health care, and electricity. "How can one seriously suggest that the village kid in India should give up her hopes of prosperity, education, and health care today, in order to prevent rising ocean levels many years down the road? What would Americans do in the same situation? Or Europeans? Or human beings anywhere?"[40]

The iron law of climate policy holds everywhere around the world, in rich and poor countries. For instance, in coming years the United

Kingdom faces the prospects of an energy shortage due to the closing both of coal plants (in turn due to laws governing their particulate emissions) and of nuclear power plants (as part of a long-term plan to reduce dependence on nuclear power), leaving few short-term options to meet expected demands for power. Possible measures to increase energy supply include building more gas-fired plants (which risks a greater dependence on Russian gas and all of the accompanying insecurities), building new nuclear plants or putting off closure of existing plants (despite significant public opposition), and building new, cleaner coal plants (despite their carbon footprints).[41] Of the choice, a UK government official explained that in "a decision between building a new coal plant and letting the lights go out—that's a no-brainer." The *Economist* interpreted that comment to signify that "something has to give, and it will probably be environmental targets."[42]

The high prices of oil in the summer of 2008 provided a real-world test of how the global public responds to significantly higher costs of energy. *The Guardian* provided a quick tour of reactions around the world:

> Concerns were growing last night over a summer of coordinated European fuel protests after tens of thousands of Spanish truckers blocked roads and the French border, sparking similar action in Portugal and France, while unions across Europe prepared fresh action over the rising price of petrol and diesel. . . . Protests at rising fuel prices are not confined to Europe. A succession of developing countries have provoked public outcry by ordering fuel-price increases. Yesterday Indian police forcibly dispersed hundreds of protesters in Kashmir who were angry at a 10 percent rise introduced last week. Protests appeared likely to spread to neighboring Nepal after its government yesterday announced a 25 percent rise in fuel prices. Truckers in South Korea have vowed strike action over the high cost of diesel. Taiwan, Sri Lanka and Indonesia have all raised pump prices. Malaysia's decision last week to increase prices generated such public fury that the government moved yesterday to trim ministers' allowances to appease the public.[43]

In the face of such political realities, policy makers find themselves conflicted, but they are not confused. They are conflicted because they express a desire to increase the costs of energy while simultaneously expressing a desire to lower those costs. At the same time they are not confused, because when such a trade-off is made, it is inevitably made in the direction of sustaining economic growth. Gwyn Prins, of the London School of Economics, called the contortions of policy makers on energy policy "gloriously incoherent" after observing their behavior at preparatory meetings immediately preceding the 2008 G8 Summit in Toyako, Japan.[44] In a morning session, Prins relates, policy makers discussed ways to lower the costs of gasoline brought on by the massive run-up in oil prices in 2007 and 2008. They broke for a nice lunch and then in the afternoon reconvened to consider ways to increase the costs of gasoline through caps or taxes in order to address ever-growing greenhouse gas emissions around the world.

A poll in the United States conducted in the summer of 2009 helpfully illustrates the iron law of climate policy. The poll asked respondents about their willingness to support a climate bill in the U.S. Congress at three different annual costs per household. At $80 per year a majority said that they would support a bill (see Figure 2.4). But at $175 per year support dropped by almost half, with a majority expressing opposition to such a bill. At $770 per year opposition exceeds support by a ratio of about ten to one. Some might argue that the poll indicates that the environment and the economy are in fact necessary trade-offs. This would be an incorrect reading of the poll. What the poll shows is that when the environment and economy are presented as trade-offs, the economy invariably wins. The implication is not that such a trade-off is inevitable; rather, any policy that seeks to achieve an accelerated decarbonization of the global economy must be designed such that economic growth and environmental progress go hand in hand.

Savvy politicians get the iron law of climate policy. Al Gore, for instance, has advocated the U.S. climate bill on the basis that it would cost the American household about "a postage stamp a day."[45] The message, of course, is that it won't affect anyone's wallet in any significant

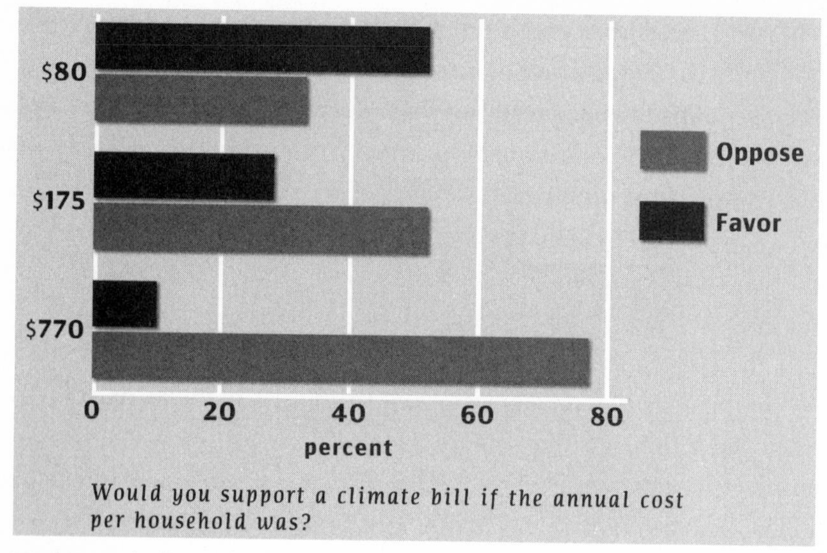

FIGURE 2.4 An illustration of the "iron law" of climate policy. Source: Poll commissioned by *The Economist*.

way. In an early 2009 debate over cap-and-trade legislation House Speaker Nancy Pelosi (D-CA) argued that "there should be no cost to the consumer."[46] Similarly, Senator Barbara Boxer (D-CA) promised that "any kind of cap-and-trade system that comes forward will not raise energy and gas prices."[47] And to remove any doubt about their intentions, the U.S. Senate voted 98–0 in the first half of 2009 to express its intention that climate legislation would not increase the tax burden on U.S. citizens.[48] As Figure 2.4 shows above, people are willing to pay some price for policies focused on climate goals, but their willingness has limits. These limits exist at different thresholds around the world.

The iron law of climate policy says that even if people are willing to bear some costs to reduce emissions, they are willing to go only so far. The iron law holds in rich countries and in poor countries, and will continue to hold. What this means is that to succeed, any policies focused on decarbonizing economies will necessarily have to offer short-term benefits that are in some manner proportional to the short-term costs. Ultimately, this means that action to achieve environmental goals will have to be fully compatible with the desire of people around the world to meet economic goals. There will be no other way.

Do We Have All the Technology We Need?

Is decarbonizing the global economy simply a matter of deploying existing technologies? According to Al Gore, it is: "We have at our fingertips all of the tools we need to solve three or four climate crises—and we only need to solve one."[49] Rajendra Pachauri has offered a similar optimistic message about technology: "All the technologies that are required for moving on a path of stringent mitigation are available to us, or on the verge of being commercialized."[50] The message that we have all the technology that we need is the flip side of the assertion that all we lack is political will. But as was the case with political will, assertions that we have all the technology we need (or soon will) simply do not square with the numbers in any practical sense.

The view that decarbonization of the global economy is a political problem and not a technological problem has been strongly influenced by a 2004 analysis by two Princeton researchers, Stephen Pacala and Robert Socolow, that was published in *Science*.[51] The analysis is often referred to by its very useful focus on a concept called a "stabilization wedge." The stabilization wedge approach, like any analysis, begins with assumptions. In this case the key assumptions were that global emissions should average 7 GtC of carbon (i.e., 2004 levels) from 2005 to 2054 to be consistent with a stabilization target of 500 ppm of carbon dioxide in the atmosphere.[52] The approach assumed that global emissions of carbon dioxide would increase from 2005 to 2054 by an annual rate of 1.5 percent, so from 7 Gt in 2004 to about 14 Gt in 2054. To average 7 Gt would require that cumulative emissions over the period 2005 to 2054 be reduced from the baseline projection by more than 160 Gt, or about twenty-three years' worth of global carbon dioxide emissions in 2004.

To get a conceptual handle on this enormous challenge Pacala and Socolow introduced the concept of a "stabilization wedge." Each wedge simply represented a cumulative 25 Gt of carbon over the period 2005 to 2054 (equivalent to 91.7 Gt of carbon dioxide). They then asked: what technologies do we now have that reduce emissions by a single wedge over that period? In their paper, Pacala and Socolow identified fifteen possible stabilization wedges, including approaches such as carbon capture and sequestration (CCS) from coal power plants, enhanced nuclear power, and improved soil management in agriculture. They argued that only seven of these fifteen approaches would have to be implemented to meet the trajectory of emissions that they propose. The implication was that there are more than enough technologies and approaches available to, in their words, "solve the climate problem for the next fifty years."[53]

The fundamental flaw in the stabilization-wedge approach was identified almost immediately by physicist Marty Hoffert of New York University, who explained that the assumption of 1.5 percent annual growth in emissions depended upon those emissions occurring at a rate considerably lower than aggregate economic growth.[54] Hoffert explained that carbon emissions were in fact growing at a rate much higher than 1.5 percent per year, "as the world switches to coal because of gas and oil production peaking, as energy demand keeps rising driven by the economic growth of China and India. The result is that carbon emissions are now growing at three percent per year, not 1.5 percent per year."[55] This meant that Pacala and Socolow had built into their analysis assumed improvements in efficiency and decarbonization of energy supply that were not being explicitly counted. In other words, many more wedges were needed than they had admitted; instead, those wedges were hidden in their assumptions.

In a subsequent 2006 paper, following Hoffert's criticisms, Pacala and Socolow expanded on the concept of stabilization wedges and acknowledged that their conclusions in fact depended upon the presence of eleven additional "virtual wedges," as shown in Figure 2.5. Carbon dioxide emissions growing at a rate lower than economic growth implies a background spontaneous decarbonization of the economy. However, even if spontaneous, Hoffert explains that "'virtual wedges imply real technology!"[56] With respect to the total of eighteen wedges needed under the full analysis, Hoffert concluded that it is therefore obvious that all of the fifteen technologies presented by Pacala and Socolow in their original analysis, even if fully deployed, would be insufficient to meet the challenge of decarbonizing the global economy in a manner

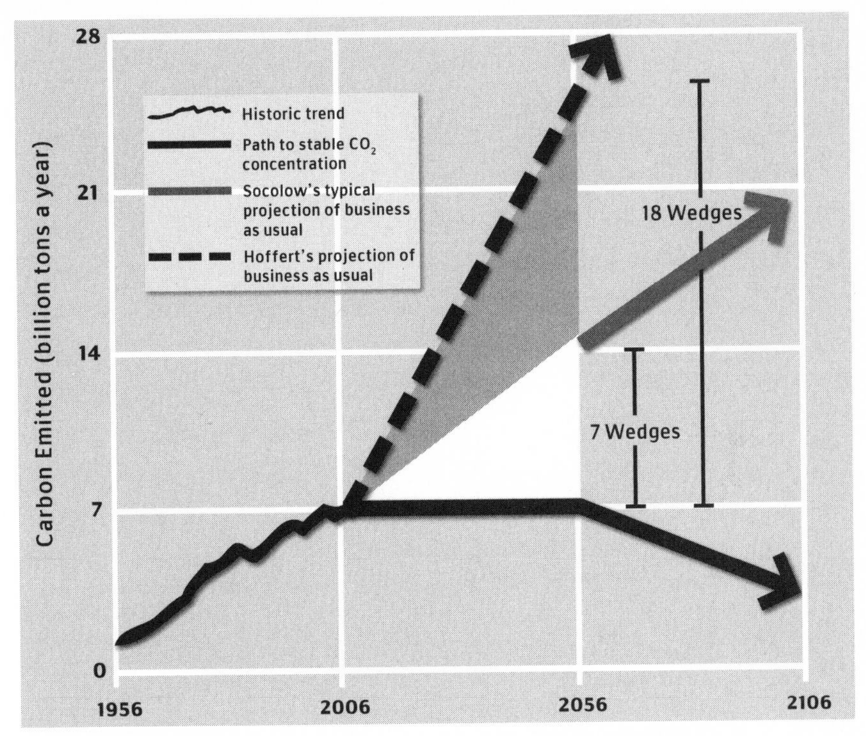

FIGURE 2.5 Virtual stabilization wedges. Source: The Breakthrough Institute.

consistent with a stabilization target of 500 ppm. Hoffert argued that the situation is even worse, as the number of wedges "explodes after the mid-century mark, if world GDP [gross domestic product] continues to grow three percent per year."[57]

The shortfall in the number of needed stabilization wedges is even more profound than Hoffert argued. In their paper Pacala and Socolow assumed that the oceans would continue to absorb carbon dioxide, as they have in the past, thus doing a large amount of work in offsetting accumulating carbon dioxide in the atmosphere. While the ocean uptake would be good news for the atmosphere, it would be potentially very bad news for the oceans, as it would alter their chemistry with unknown consequences. Pacala and Socolow assumed that 180 Gt of carbon would be absorbed by the oceans from 2005 to 2054, representing more than seven wedges' worth of emissions, and in their discussion of uncertainties figured that this value could be even higher.[58]

So under the assumptions of Pacala and Socolow, stabilizing emissions (not concentrations) at 2004 levels within fifty-four years would require the seven wedges that they explicitly identified in the *Science* paper, the eleven "virtual wedges" that are implied by spontaneous decarbonization that they later acknowledge, plus additional wedges to deal with the oceanic uptake of carbon dioxide (often called ocean acidification), for a total of twenty-five or more wedges. This means that even full deployment of all fifteen of the options proposed by Pacala and Socolow could leave as much as half the job remaining. To put this into context, many calls for action seek to limit concentrations to 450 parts per million, and some even 350 ppm, much less than the 500 ppm assumed in the Pacala and Socolow analysis. These lower targets require cutting emissions by 50 percent or more from 2010 levels by 2050. Achieving such targets implies even more wedges.

Looking at Table 2.1, and assuming that all of these massive steps implied by these fifteen categories of action are implemented, where are up to fifteen additional wedges going to come from? Even full implementation of each of the fifteen categories stretches credulity. The real lesson to take from the stabilization wedge analysis is a practical one: we simply do not have all the technology to allow for realistic deployment and displacement of existing infrastructure.[59] Technology must advance if we are to make rapid progress at decarbonizing the global economy.

The short history of global emissions and economic growth since the 2004 publication of Pacala and Socolow's paper shows that Hoffert was absolutely correct in his critique. Emissions in 2005 and 2006 increased by an average of more than 3.1 percent per year.[60] Even in 2008, during the global financial crisis, emissions increased by 1.7 percent.[61] In 2009, after taking into account the global financial crisis and economic slowdown of the previous year, the U.S. Energy Information Agency (EIA) projected in 2010 that global emissions would top 11 Gt of carbon by 2030, five years (and two wedges) ahead of the pace assumed in the stabilization-wedge scenario.[62]

In 2008, along with Tom Wigley and Chris Green, I published a paper in *Nature* asserting that the IPCC had likely underestimated the challenge

TABLE 2.1 Fifteen potential stabilization wedges: Strategies available to reduce the carbon emission rate in 2054 by 1 Gt of carbon/year or to reduce carbon emissions from 2004 to 2054 by 25 Gt of carbon

Option	*Effort by 2054 for one wedge, relative to 14 Gt of carbon/year business as usual*
1. Efficient vehicles	Increase fuel economy for 2 billion cars from 30 to 60 mpg
2. Reduced use of vehicles	Decrease car travel for 2 billion 30-mpg cars from 10,000 to 5,000 miles per year
3. Efficient buildings	Cut carbon emissions by one-fourth in buildings and appliances projected for 2054
4. Efficient coal plants	Produce twice today's coal-power output at 60 percent instead of 40 percent efficiency (compared with 32 percent today)
Fuel shift	
5. Gas power for coal power	Replace 1,400 coal plants with gas plants (four times the current production of gas-based power)
Carbon dioxide capture and storage	
6. Capture carbon dioxide at power plant	Introduce CCS at 800 GW coal or 1,600 GW natural gas
7. Capture carbon dioxide at hydrogen-producing plant	Introduce CCS at plants producing 250 megatons of hydrogen per year from coal or 500 megatons hydrogen per year from natural gas (compared with 40 megatons hydrogen per year today from all sources)
8. Capture carbon dioxide at coal-to-synfuels plant	Introduce CCS at synfuels plants producing 30 million barrels a day from coal, if half of feedstock carbon is available for capture
Geological storage	Create 3,500 Sleipners (so-called after a Norwegian gas plant that captures carbon in geological formations)
Nuclear fission	
9. Nuclear power for coal power	Add 700 GW
Renewable electricity and fuels	
10. Wind power for coal power	Add 2 million 1-MW-peak windmills
11. PV power for coal power	Add 2,000-GW-peak PV
12. Wind-produced hydrogen in fuel-cell cars for gasoline in hybrid car	Add 4 million 1-MW-peak windmills
13. Biomass fuel for fossil fuel	Add 100 times 2004 Brazil or U.S. ethanol production, with the use of one-sixth of world cropland

(continues)

TABLE 2.1 (continued)

Forests and agricultural soils	
14. Reduced deforestation, plus reforestation, afforestation, and new plantations	Decrease tropical deforestation to zero and establish 300 million hectares of new tree plantations
15. Conservation tillage	Apply to all cropland (ten times 2004 usage)

Source: Modified from S. Pacala and R. H. Socolow, "Stabilization Wedges: Solving the Climate Problem for the Next 50 Years with Current Technologies," *Science* 305 (2004): 968–972.

of reducing emissions. We argued that the IPCC had built into its scenarios of future emissions historically unprecedented rates of spontaneous decarbonization, just as had been done in the case of the stabilization wedges.[63] To paraphrase Marty Hoffert, spontaneous decarbonization implies real actions. We pointed out that of these presumed real actions, "the IPCC assumes that the majority of the challenge (between 57 percent and 96 percent) of achieving stabilization at around 500 parts per million will occur automatically, leaving a much smaller emissions reduction target for explicit climate policies." With the relatively small amount of action remaining, after the spontaneous decarbonization was factored in, the IPCC concluded that stabilizing concentrations of carbon dioxide levels in the atmosphere "can be achieved by deployment of a portfolio of technologies that are currently available and those that are expected to be commercialised in coming decades."[64]

Time will tell whether the IPCC was correct in its assumptions that accelerating decarbonization of the global economy requires few policies in comparison to those actions that will occur automatically. However, the debate is of more than just academic interest, as assumptions shape policies, especially policies focused on stimulating innovation. On the one hand, if one believes, as the IPCC does, that the majority of decarbonization of the global economy will happen spontaneously in the absence of explicit policies, then one would be less apt to support investments in technological innovation that seeks to bring down the costs of alternatives to fossil fuels or to surmount other obstacles to widespread deployment of low-carbon energy. On the other hand, if one believes that it is plausible that the decarbonization challenge has been understated, then it would be prudent, even necessary, to focus more attention and resources on the need for innovation in energy technology as a core decarbonization policy. Our 2008 paper concluded with such a warning: "The IPCC plays a risky game in assuming that spontaneous advances in technological innovation will carry most of the burden of achieving future emissions reductions, rather than focusing on creating the conditions for such innovations to occur."

The stabilization wedges and the IPCC have shaped the policy debate on decarbonization away from technological innovation, under an assumption that we have all the technologies that we need (or soon will have them). In a very practical sense, that assumption is very likely to be wrong, a point that will be repeated and emphasized in quantitative fashion at the end of Chapter 4.

Like so much in the climate debate, there is more to the story. In a 2008 interview Stephen Pacala made a remarkable admission. The purpose of the stabilization-wedge paper, he said, was in fact entirely political. Pacala explained that the paper was written to counter the influence in the policy debate of the work of Marty Hoffert and his colleagues, who (as we have seen) argued persuasively that we currently did not have the technologies that we needed to decarbonize the global economy in a manner consistent with low stabilization targets. The problem, Pacala said, was not that Hoffert's work was necessarily wrong. The problem was that Hoffert's work was being taken seriously in the political process by opponents to action, and Pacala explained that he wanted to counter this influence: "The purpose of the stabilization wedges paper was narrow and simple—we wanted to stop the Bush administration from what we saw as a strategy to stall action on global warming by claiming that we lacked the technology to tackle it."[65] In setting forth a view that we did in fact have all the technologies that we needed, Pacala said that the stabilization-wedge paper was "surprisingly effective." On the political motivation behind the paper, Pacala justified it as follows: "That doesn't mean that there aren't things wrong with it and that history won't prove it false. It would be astonishing if it weren't false in many ways, but what we said was accurate at the time." The influence of the stabilization-wedge paper continues to shape the climate debate, long after it has been shown to be seriously flawed, and long after George W. Bush retired to Texas.

Apparently, Pacala and Socolow decided to fight a political battle through science by seeking to take away a justification used by the Bush administration to emphasize technologies as a core of their climate policy response. To a surprising degree they were successful in countering that justification. However, although the Bush administration's motives may have been to forestall other action on climate, the world does in fact need investments in innovation if there is to be any hope of stabilizing atmospheric concentrations of carbon dioxide at low levels. Pacala and Socolow's paper won a short-term political battle, but only at great cost to the longer-term challenge of decarbonization. Unfortunately, we do not have all the technology that we need. Any commonsense climate policy will take a look at the real numbers behind the stabilization wedges and recognize that technological innovation must be a central strategy behind any effective policy focused on accelerating decarbonization.

Ground Rules for a Commonsense Climate Policy

The three fallacies examined in this chapter together form a set of boundary conditions for policies focused on achieving emissions reductions, or other goals, under the notion of climate policy. As theologian Reinhold Niebuhr might have put it, effective climate policy will result only from an appreciation of those conditions in society that are extremely difficult to change via advocacy or policy and those that are more amenable to change, and the wisdom to be able to tell the difference between the two.

This chapter's overview of public opinion on climate change over the past several decades in the United States and internationally indicates that there is ample political will for action. In fact, public opinion is as strong (or stronger) for action on climate change as it has been on many public issues environmental or otherwise for which substantial action has occurred. Many of the claims of a lack of political will are simply circular arguments, rather than a careful analysis of how public opinion on climate change actually compares in historical context to other issues for which action actually occurred. Such context strongly suggests that public opinion on climate change is strong enough to support action. Effective action will be shaped around the realities of public opinion, rather than unrealistic expectations about shaping public opinion around theoretical or technocratic policies invented by policy wonks. The test to apply is not whether a proposed policy works in theory but if it does in practice. The question that follows is, of course, what policy?

A related issue is that many advocates for action have emphasized that climate policy requires sacrifice, as economic growth and environmental progress are necessarily incompatible with one another. This perspective has even been built into the scenarios of the IPCC and is advocated by its chairman. However, experience shows quite clearly that when environmental and economic objectives are placed into opposition with one another in public or political forums, it is the economic goals that win out. I call this the iron law of climate policy. Opinion polls show that the public is indeed willing to pay some amount for attaining environmental goals, just as it is with respect to other societal goals. However, the public has its limits as to how much it is willing to pay. What this means is that climate policies must be made compatible with economic growth as a precondition for their success.

Some may wail and scream about this fundamental reality, and instead demand that people around the world reorder their values such that environmental objectives trump considerations of economic growth. Considerable effort has been expended by advocates on many issues in an effort to reshape societal values. The world's religions, for example, have been trying to reorder values for millennia. While over time value reshaping certainly occurs, there is little evidence to suggest that efforts to achieve a global value restructuring offer a useful path forward on decarbonization of the global economy, or improved resilience to climate extremes. Of course, people will still try. Even so, for climate policy the reality is that the iron law of climate policy will hold fast for the foreseeable future. Those interested in real progress on decarbonization of the global economy should respond accordingly.

A third boundary condition is that existing technologies are not in any practical sense up to the task of decarbonizing the global economy.

Sure, one can invent fanciful scenarios that involve extreme global depopulation or economic collapse that leads to dramatic emissions reductions. One can even invent seemingly more realistic scenarios involving the deployment of a new nuclear power station somewhere in the world every day for the next four decades or the deployment of 2 million wind turbines (see Chapter 4), but these are not practically realistic scenarios. The fact is that no one knows how to decarbonize a large economy, much less the world, using existing technology on timescales implied by emissions-reduction targets currently suggested by policy makers. Throwing everything we can think of (for example, see again Table 2.1) at the problem is not nearly enough.

A review of "what we know for sure, but just ain't so" provides a few boundary conditions that suggest design criteria for any successful policy focused on decarbonization of the global economy:

1. Climate policies should flow with the current of public opinion rather than against it.
2. Efforts to sell the public on policies that will create short-term economic discomfort cannot succeed if that discomfort is perceived to be too great. The greater the discomfort, the greater the chances of policy failure. Short-term costs must be commensurate with short-term benefits.
3. Innovation in energy technology—related both to the production of energy and to its consumption—necessarily must be at the center of any effort to accelerate decarbonization of the global economy.

I'll return to these design criteria in Chapter 9 when I outline a perspective on how climate policy might be different, and perhaps more likely to show progress. The next two chapters will reinforce these design criteria by looking at the simple mathematics of real-world challenges of decarbonizing the global economy. Wisdom in policy analysis begins with a clear-eyed view of the scope of a challenge, and that is where we turn next.

CHAPTER 3

Decarbonization of the Global Economy

As explained in Chapter 1, the accumulation of carbon dioxide in the atmosphere influences the climate, changes the chemistry of the oceans and causes them to rise, and influences the growth of vegetation, among other things. All of these effects lead to further effects in the earth system. Some of these effects may be predictable, and some may not; some could be benign, or even beneficial, but others might be far less acceptable. Much attention in the field of climate science in recent decades has been focused on trying to gain a clearer view of the future impacts of accumulating carbon dioxide (and other greenhouse gases, as well as other natural and human influences). The literature on this topic is so vast that one could easily cherry-pick a few studies suggesting that the impacts may be benign or, in contrast, that those impacts may be catastrophic. Science cannot presently adjudicate between these possibilities, or even give reliable odds on particular outcomes, leaving what Steve Schneider calls a "lingering frustration" (see Chapter 1). Many, if not most, scientists believe that the impacts will be on balance negative and significant.

For some people the mere fact that significant negative impacts are possible is all they need to know to support policies focused on accelerating decarbonization of the global economy. Undoubtedly in some cases, the issue of climate change simply adds weight to actions that they would support for other reasons, such as expanding alternative energy or redistributing wealth. But for others the science by itself provides an insufficient basis for action, especially costly and aggressive action. Here as well science is filtered through preexisting views and commitments. Climate change is a bit like a policy inkblot on which people map onto the issue their hopes and values associated with their vision for what a better world would look like. In such a circumstance it should not be a surprise that scientific information cannot lead to political consensus.

Even so, the political battle over climate change has been waged through science. Advocates for action typically seek to compel the recalcitrant by offering up ever more certain scenarios of an apocalyptic future. Those seeking to prevent action highlight the uncertainties in climate science, some going so far as to call the issue a myth or a hoax. Some of those on each side of this debate have misrepresented climate science in political debates. For those advocating action, this strategy has failed; instead, the strategy has contributed to a problematic politicization of climate science and increased opposition to action. As argued in more depth in Chapter 7, waging a political battle through science tilts the playing field in the direction of those opposing action and threatens the integrity of climate science as well.

But science need not carry all of the weight of advocacy for action to accelerate decarbonization of the global economy, as there are other, far less controversial, reasons that decarbonization makes sense. Together these other reasons may not be sufficient to justify decarbonization to levels implied by very low targets for stabilizing concentrations of carbon dioxide, but they are certainly strong enough to motivate the first steps on that path, which so far have been very difficult to take. And after the first steps are taken, the ones that follow might then come a bit easier.

The World Needs Vastly More Energy

Energy experts use a dizzying array of units and jargon to talk about energy. In this chapter and the next I use some back-of-the-envelope approximations to help make better intuitive sense of energy and its relationship with carbon dioxide emissions. I also use a standard set of units. One of these units is the "quad"—a useful concept for discussing the consumption of energy. The term "quad" is shorthand for 1 quadrillion (1,000,000,000,000,000) British thermal units. A Btu refers to the amount of energy required to elevate the temperature of one pound of water by one degree Fahrenheit.[1] But a more intuitive conversion is to think about a quad in terms of power plant–equivalent electricity generation. One quad is equivalent to about 11 gigawatts (GW) of electricity (over one year).[2] How much is 11 gigawatts? It is the amount of electricity produced by about fifteen typical power plants, each generating 750 megawatts (MW) of electricity.[3] In recent years the United States as a whole consumed about 100 quads of energy each year.[4]

In a 2009 report the U.S. Energy Information Agency estimated that the world would consume 508 quads of energy in 2010. The EIA estimated that by 2030 the world would consume a total of 678 quads of energy, which represents a growth rate of about 1.5 percent per year in the context of global economic growth expected to be perhaps twice as great. Thus, the EIA scenario for future energy consumption already factors in aggressive improvements in energy efficiency. Consider that if demand were to increase by 2 percent annually to 2030 (instead of 1.5 percent), the world would need an additional 77 quads in 2030, for a total of 755 quads. To reach this level of total energy consumption would be the equivalent of adding more than 3,700 new power plants![5] And the demand for energy is likely to continue to increase well beyond 2030.

The precise total amount of energy that the world needs in coming decades will be determined by how fast the global economy grows and the nature of that economy, including the global mix of activities (e.g., services versus manufacturing) and the efficiency with which those activities are conducted. As the simple exercise above shows, small differences in any of the variables that shape the economy and its use of energy can lead to dramatically divergent outcomes over decades. It is for this reason that Vaclav Smil of the University of Manitoba summarizes the track record of energy forecasting as follows: "With rare exceptions, medium- and long-range forecasts become largely worthless in a matter of years, often just a few months after their publication."[6] Smil recommends "contingency scenarios" to explore "what ifs" and "no-regret normative scenarios" that shape a course in a desired political direction.

From the what-if perspective, it is essential to realize that under all plausible scenarios, in the coming decades the world is going to need more energy—vastly more energy. Meeting the increasing global demand will be facilitated by an increasing diversification of energy supply beyond coal, gas, and oil. Wherever one falls on the spectrum of debate about peak energy supply—that is, the debate over the finite nature of fossil fuels (oil, gas, and coal) and when production will peak and then decline—it is clear that the demand for a robust global energy supply to meet ever-increasing demand will be largely insensitive to the point, if and when it is reached, of peak production. Beyond hydrocarbons there is also debate about peak uranium to supply nuclear reactors and peak rare-earth minerals such as dysprosium and terbium, used in technologies such as wind turbines and energy-efficient lightbulbs.[7] Uncertainties about the future marginal costs of various forms of energy supply support the need for a more robust energy supply. And having a robust energy supply thus means diversification. In many respects (but, as will be seen, not all), diversification of supply means accelerating the pace of decarbonization of the global economy.

Energy Dependence Exacerbates Insecurities

It is not an overstatement to observe that the benefits of contemporary modern society are due in large part to cheap energy supply from oil, gas, and coal. An energy-industry venture capitalist guest lecturing in one of my courses once commented on the "near-miraculous" feat of bringing a gallon of gasoline to your car, distilled from crude oil from thousands of feet below the ocean floor, enabling near-unlimited mobility at a price of only a few dollars per gallon. And he was right. The modern energy economy is a testament to human know-how and ingenuity. But is it possible to both celebrate the accomplishments of a world built on carbon dioxide–emitting sources of fuel and recognize that there are reasons to look forward to a future with a different energy mix than the one that has brought us to today?

One important issue facing nations around the world is energy security, which can refer to both security of supply and the security that results when energy is supplied reliably and at low cost. For instance, Raja Pervez Ashraf, Pakistan's minister for water and power, commented in 2009 that "Pakistan has to make a choice whether to develop electricity or face power cuts that result in unemployment, low economic growth, and protests."[8] He views securing access to energy as central to enhancing Pakistan's domestic security. The view from Pakistan is no different from the view from Africa, Southeast Asia, or other locales where energy supply is neither readily available nor inexpensive.

Consider that in 2008 approximately 1.5 billion people worldwide lacked access to electricity. About 600 million of these people were in sub-Saharan Africa and 800 million in Asia. The International Energy Agency (IEA) explains why access to electricity matters: "It is impossible to operate a factory, run a shop, grow crops, or deliver goods to consumers without using some form of energy. Access to electricity is particularly crucial to human development as electricity is, in practice, indispensable for certain basic activities, such as lighting, refrigeration and the running of household appliances, and cannot easily be replaced by other forms of energy."[9] The lack of access to electricity helps explain why it is that countries with large and poor populations provide little support for efforts to reduce emissions of carbon dioxide if such efforts imply any extra costs. In October 2009 during the lead-up to the December Copenhagen climate conference, Indian prime minister Manmohan Singh explained simply that "developing countries cannot and will not compromise on development."[10] This, of course, is another invocation of the iron law of climate policy.

In 2009 the IEA published an aggressive emissions-reduction scenario, consistent with ambitious targets for stabilizing concentrations of carbon dioxide in the atmosphere as called for by many environmental campaigners. Incredibly, the scenario, which if followed would no doubt be greeted as a "success" by many campaigners for action on climate change, reduced the number of people worldwide without access to electricity by less than 14 percent from 2008 levels, leaving 1.3 billion people in the dark.[11] To the extent that the IEA scenario is broadly representative of the climate policies of developed nations, the scenario represents a total refusal on those countries' part to countenance the circumstances facing developing ones. Connie Hedegaard, Denmark's energy and climate minister and host of the 2009 Copenhagen climate meeting, expressed this view when she explained with respect to the need for developing countries to reduce their emissions that "China and other emerging nations must accept it even if it isn't fair."[12]

Here Hedegaard runs smack into the iron law of climate policy. As we have seen, if development is viewed as a trade-off of emissions reductions, then development will always win out. Creating a climate policy in which development and emissions reductions go hand in hand thus far has not been a focus of climate policy, regardless of the rhetoric of policy debate. If it were, then increasing access to energy via a diversification of supply would be a much more prominent feature of policy proposals, and organizations like the IEA would not advance scenarios with more than a billion people still lacking access to electricity in 2030. An approach focused on expanding access to energy while also diversifying supply would almost certainly be better received by developing countries than one that implicitly or explicitly questions their desire for continued economic growth.

Energy is necessary for development, but it is also, thanks to its cost, an obstacle to the same. Author and journalist Robert Bryce estimates in back-of-the-envelope fashion that the direct costs of fuel alone resulted in about $5 trillion of expenditures in 2008, which is about 8 percent of global GDP.[13] Bryce's estimate assumed a price of oil of $60 per barrel. At two or three times that value, the proportion of global GDP devoted to energy increases in similar fashion. The trillions of dollars spent meeting basic energy needs are not available for investments in education, health, and other important aspects of development. Improving access and security of the supply of energy will necessitate reducing the costs of energy relative to global GDP over the long run, a tall order given the energetic punch packed by fossil fuels and the prospects of their limited supply. Innovation in energy technologies offers the promise of lower costs, and thus prospects for increased access and supply. The IEA has suggested that the world will need to invest more than $500 billion per year until 2030 to transform the global energy system. This number seems large, and of course it is in an absolute sense. But relatively it is not so large, representing an added cost of only about $6 per barrel of oil.[14] Whether the actual investment needed is larger or smaller than that suggested by the IEA, the number does suggest a level of effort comparable in scope to the U.S. military budget during the years of the cold war.

The high costs of energy have tangible, real-world effects today. The UN's Food and Agricultural Organization explained that the high costs of food in 2009, exacerbated by the global financial crisis, contributed to an increase in the number of undernourished people around the world to the highest levels since 1970, with more than 1 billion people classified as undernourished.[15] The FAO argued that, because the energy market is so much bigger than the grain market, energy prices may be as or more important for determining the cost of food than the food supply is.[16] Securing the energy supply, and with it certainty in cost and access, has a key role in dealing with the global challenges of food security and malnutrition. From this perspective, it can easily be seen why biofuels based on food grains can serve to undermine food security when they are in economic competition with food. This is one reason advocates of biofuels have increased their attention to those plants that are not in direct competition with food.

It is not just poor countries that are sensitive to issues of energy security. In the winters of 2006 and 2008–2009, Russia shut off gas deliveries to eastern Europe during a dispute with Ukraine.[17] Because Europe receives a considerable amount of gas from Russia, effects on gas supplies were felt as far west as France.[18] An EU spokesperson complained, "It is unacceptable that the EU gas supply security is taken hostage to negotiations between Russia and Ukraine."[19] Ironically enough, according to the *Financial Times*, the expansion of wind power in western Europe exacerbated its dependence on gas from the East, because gas-fired power plants are needed during periods of low wind.[20] Efforts to diversify supply can have counterintuitive consequences.

It is also important to observe that security of supply is not always consistent with efforts to decarbonize. For instance, the United States has vast reserves of coal, which could provide an opportunity to reduce energy dependence, but at the same time would also increase carbon dioxide emissions. In order for coal-powered electricity to be compatible with goals of rapid decarbonization, it would be necessary to develop and deploy technologies to capture and store carbon dioxide emitted from power plants. There is simply no alternative. A different sort of trade-off is implied by nuclear power, which offers an appealing path toward diversifying energy supply while not increasing carbon dioxide emissions. But nuclear power is also the subject of intense domestic and international concern for reasons of security, in this case not associated with energy supply but the risks of nuclear power, waste, and associated nuclear technologies falling into the hands of those with bad intentions. All technologies of energy supply face trade-offs among various competing interests. Consequently, increasing energy security thus involves balancing a range of concerns. For many countries without significant oil, gas, and coal supplies, diversifying supply may in fact mean a move away from these sources to ones that can be sourced locally. The extent to which such diversification proves feasible will depend a great deal on the cost of alternative energy supply, with less expensive alternatives to fossil fuels aiding efforts to achieve greater diversification.

But perhaps the most compelling reason to accelerate decarbonization of the global economy is that, as discussed below, the world has already been decarbonizing for more than a century. Decarbonization—defined in Chapter 1 as the process of growing the economy at a rate faster than the rate of growth in carbon dioxide emissions—has historically been associated largely with increased efficiency in the use of energy, and to a somewhat lesser degree in the decarbonization in the energy supply. Efforts to secure a diverse energy supply and to improve the efficiency of energy use together provide a compelling reason to at least get started on the challenge of accelerating the decarbonization of the global economy. Whether such justifications are sufficient to carry an effort forward to 2050 is uncertain (keep in mind the pitfalls of energy forecasting), but they are sufficient to encourage building a broad coalition for starting the job now. The push to improve global energy policies thus has both climate- and non-climate-related justifications; together they provide a broad footing for making the case for accelerating the decarbonization of the economy. So far debate over climate policies has focused too much on climate and too little on the benefits of diversification of, access to, and costs of the energy supply.

The remainder of this chapter will argue that decarbonizing the global economy is an enormous task, requiring a much more direct approach than most national and international policies on carbon dioxide emissions have countenanced. A direct approach necessarily focuses explicitly on improving energy efficiency and decarbonizing the energy supply. In fact, it is only through these mechanisms that emissions will be reduced, whether one concentrates explicitly on those mechanisms or indirectly, such as through efforts to price carbon and establish caps on emissions. The uncomfortable reality is that no one knows how fast a major economy can decarbonize, much less the entire global economy. Consequently, policy will necessarily have to proceed incrementally and experimentally, and will succeed only if the short-term benefits of action are proportional to the short-term costs. And even then efforts to stabilize carbon dioxide concentrations at a low level may not succeed, necessitating a backstop.

Decarbonization Arithmetic

Like the arithmetic of carbon dioxide concentrations we saw in Chapter 1, the arithmetic of decarbonization policies is surprisingly simple.[21] In 2000 Paul Waggoner and Jesse Ausubel wrote that to understand our ability to influence environmental outcomes through policy requires "quantifying the component forces of environmental impact and integrating them."[22] For carbon dioxide emissions there is a very simple yet powerful relationship that describes the "component forces" that together result in carbon dioxide emissions. This relationship has been called the Kaya Identity, after Japanese scholar Yoichi Kaya, director-general of the Research Institute of Innovative Technology for the Earth in Kyoto, Japan, who first proposed it in the late 1980s.

The Kaya Identity can be used to decompose the factors that lead to carbon dioxide emissions from the production and use of energy in the global economy, but also to evaluate policies aimed at reducing those emissions to a level consistent with some specific stabilization target. The Kaya Identity is composed of two primary factors. The first is economic growth (or, if the economy shrinks, contraction), which is represented in terms of GDP as a measure of the exchange of goods and service. The second factor encompasses changes in technology (or the use of technology) and includes efficiency and energy sources. Under the Kaya Identity technology is represented as carbon dioxide emissions per unit of GDP, or carbon intensity of the economy.

Each of these two primary factors—economic growth and technology—can be broken down into two further subfactors. Economic growth (or contraction) is composed of changes in population and in per capita economic activity, measured in terms of GDP. The carbon intensity of the economy is represented by the product of energy consumed per unit of GDP, called energy intensity, and the amount of carbon emitted per unit of energy, called carbon intensity. Together the four factors of the Kaya Identity—population, per capita GDP, energy intensity, and carbon intensity—explain the various influences that contribute to increasing atmospheric concentrations of carbon dioxide. Table 3.1 shows the four factors expressed as an equation.

The Kaya Identity tells us exactly what families of tools we have in the policy toolbox to reduce carbon dioxide emissions to some desired level. Specifically, carbon dioxide accumulating in the atmosphere can be reduced only by influencing the following four levers.

1. We could reduce population.
2. We could reduce per capita GDP.
3. We could become more efficient.
4. We could switch to less carbon-intensive sources of energy.

These are the four—and the only four—means of reducing carbon dioxide emissions. All policies being discussed as climate policies must influence these levers if they are to have an effect. So debates about carbon taxes, cap-and-trade programs, offsets, energy innovation, personal carbon allowances, and on and on ultimately must eventually arrive at exactly the same place.

TABLE 3.1 Kaya Identity

(1) Carbon Emissions = Population * Per Capita GDP * Energy Intensity * Carbon Intensity

a. P = Total Population
b. GDP/P = Per capita GDP

(2) GDP = Economic Growth (Contraction) = P * GDP/P = GDP

a. Energy Intensity (EI) = Total energy (TE)/GDP = TE consumption/GDP
b. Carbon Intensity (CI) = C/TE=Carbon emissions/total energy consumption

(3) Technology = "Carbon Intensity of the Economy"

= EI * CI = TE/GDP * C/TE = C/GDP

We can make the issue even simpler yet. As Chapter 2 argued, one of the ground rules of climate policies is that reducing economic growth or limiting development as a means to address increasing carbon dioxide is simply not an option—it is an iron law. Indeed, in rich and poor countries alike increasing GDP is a central focus of policy. Consider that even if per capita wealth were to stay constant, a growing global population alone implies a rising GDP and thus rising emissions.

Figure 3.1 shows another way to visualize climate policy's iron law from an analysis conducted by the United Nations of the distribution of global income for 1970 and 2000 along with an estimated curve for 2015.[23] A defining feature of the curve is its rightward movement over time, which indicates more people living at higher income levels, and also a higher global GDP. Indeed, success with respect to poverty is often measured by reducing the number of people living below some income threshold, such as $1 per day. In yet another trade-off of competing values, moving people out of poverty by increasing their incomes means increasing carbon dioxide emissions, all else staying equal. In 2005 the World Bank estimated that 95 percent of all people in developing countries live on less than $10 per day.[24] People everywhere will continue to try to increase their material well-being, and GDP will continue to be one measure of that increase. It is vitally important to note that the fact

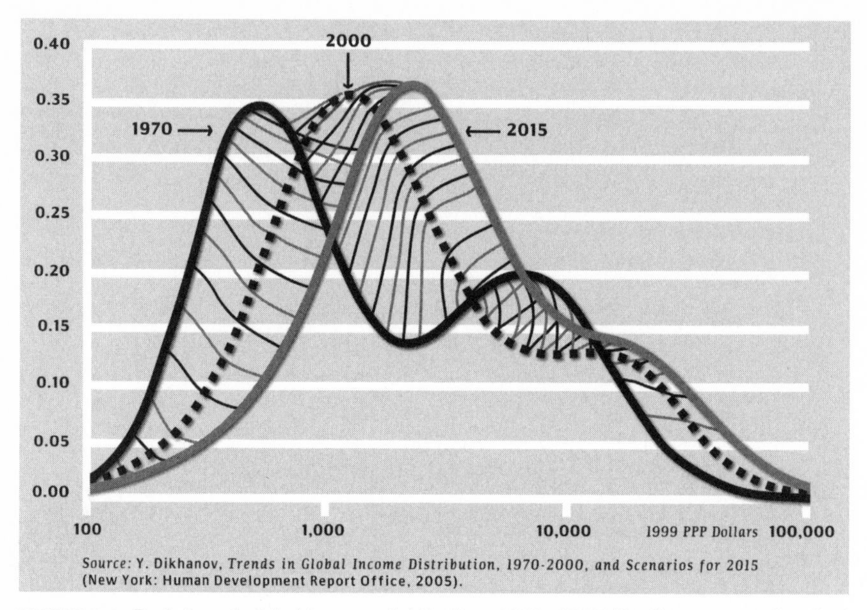

FIGURE 3.1 Evolution of global income distribution, 1970–2015 (the lines connecting the curves connect equal percentiles).

of GDP increase is quite different from the question about whether increasing GDP accurately measures what matters for human well-being. Changing how we measure and value growth won't alter the reality of emissions or its close linkage to the conventional metrics of GDP.

While it can be a useful exercise to debate the desirability of continued economic growth, its sustainability, or its expression as GDP, such debates are entirely academic from the standpoint of decarbonization. Population is expected to increase for decades, perhaps through at least midcentury, and it seems highly unlikely that policies focused on global population control (much less its managed reduction) are going to be put into place anytime soon. At the same time, people around the world are going to continue to try to improve their material standing in the world, and as they do so, an increasing GDP will result. Governments are committed to aiding their citizens in pursuing greater wealth. Consequently, climate policies will invariably be put into place in the context of a broader societal commitment to increasing GDP. As we shall see, the rate at which GDP increases makes a very big difference in how much carbon dioxide we emit into the atmosphere. So while there are those who will

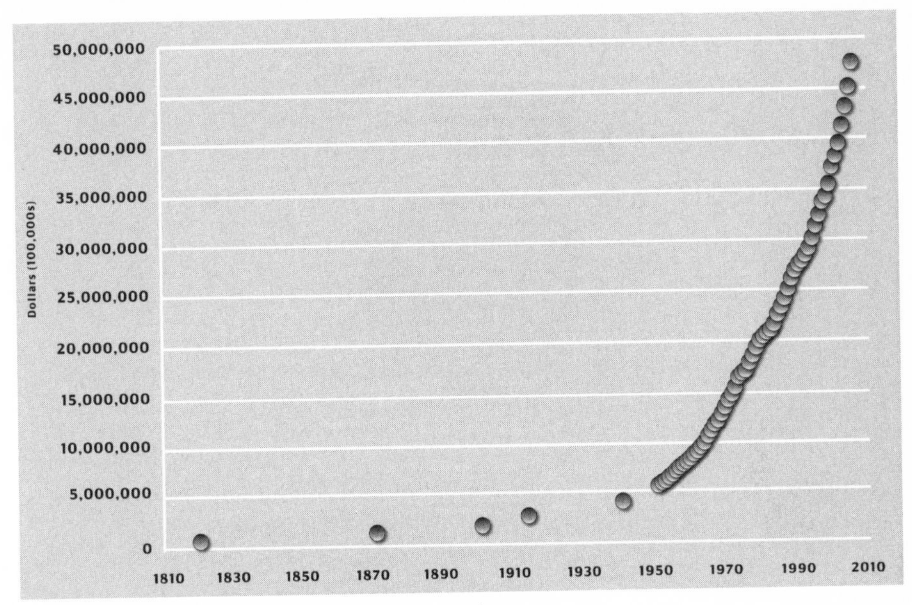

FIGURE 3.2 Global GDP, 1820–2006. Source: A. Maddison.

continue to talk about policies focused on a willful contraction or significant slowdown of global economic growth, such policies won't be further discussed in this book because they are simply not going to happen. For the foreseeable future, the iron law of climate policy is exactly that.

Figure 3.2 shows the GDP data that are used throughout the various analyses of decarbonization policies that follow in subsequent sections. The data come from research by the decorated economist Angus Maddison of the University of Groningen in the Netherlands. The data are adjusted to a common basis in U.S. dollars to allow cross-national comparisons.[25] As the figure shows, GDP growth has been sustained for the past two centuries.

All proposals advanced by governments and in international negotiations to reduce emissions of carbon dioxide focus (directly or indirectly) on actions that will lead to the reduction of the carbon intensity of the economy (whether they are explicitly presented as such or not), which is a more technical term for decarbonization. Thus, the Kaya Identity provides a straightforward and useful basis for evaluating the proposed and actual performance of policies focused on decarbonization, which are often called *mitigation* policies by those who focus on climate policy.

In 2009 the U.S. Energy Information Agency estimated total global emissions of carbon dioxide in 2010 from the combustion of fossil fuels to be about 31 Gt and projected them to rise to about 40.4 Gt by 2030.[26] The IPCC in 2007 reported a median scenario for global emissions in 2100 to be 220 Gt carbon dioxide, with its baseline scenarios ranging from 36.7 to 916.8 Gt carbon dioxide, reflecting enormous uncertainties about the future.[27] The future is unpredictable, of course, because it will (in part) be determined by the choices that we make.

Figure 3.3 shows the relationship of GDP and emissions for the period 1980 to 2006. It shows that while the relationship is not a straight line, it is pretty close. When GDP increases, so too do carbon dioxide emissions. If you take a close look at the graph you will also see that the four or five data points at the highest levels of GDP are much more spread out than the points representing earlier years (and thus lower levels of GDP). What this indicates is that during the first decade of the twenty-first century (i.e., 2000 to 2006 based on available data), the world has been *recarbonizing*: For every additional $1,000 of GDP activity, the amount of carbon dioxide generated is higher than it was during the 1990s and earlier. So even as the world's attention has been focused on climate policy, the global economy became ironically and frustratingly more carbon intensive for every additional dollar of economic activity.

In these sobering data, there is also some good news to report. Figure 3.4 shows that for about 100 years the global economy has been decarbonizing, meaning that on average globally we successively emit less carbon dioxide per unit of GDP, with values dropping by about half in 100 years, from 1.27 tonnes of carbon dioxide per $1,000 GDP in 1910 to 0.62 in 2006. This decarbonization has taken place without any attention to climate policy or explicit attention to a global decarbonization policy. Noting this trend in the 1980s and 1990s, some scholars predicted that the world was on its way to a decreasing dependence on hydrocarbon-based fuels that would result in the continued decarbonization of the global energy system during the twenty-first century.[28] Even if such scenarios come to pass, they won't help much in addressing the challenge of stabilizing concentrations of carbon dioxide at low

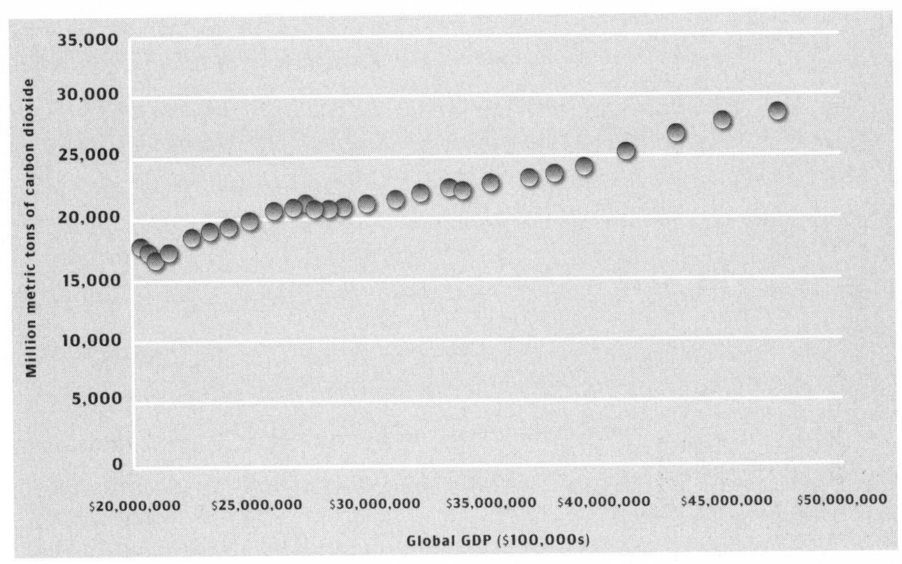

FIGURE 3.3 Relationship of GDP and emissions, 1980–2006. Source: A. Maddison and U.S. Energy Information Administration.

levels, as we shall see, because the historical rate of decarbonization is far too slow to be consistent with a low stabilization target. A considerably greater acceleration would be necessary.

Figure 3.5 shows that the primary reason for decarbonization since 1980 has been improvements in energy efficiency (i.e., energy intensity of GDP), while improvements in the carbon intensity of energy have contributed a smaller amount. If the world is going to simultaneously provide much more energy and meet aggressive targets for decarbonization, then decreases in the carbon intensity of energy supply are necessarily going to have to play a much larger role in future decarbonization than in the past. The good news is that reductions in the carbon intensity of energy can also contribute to diversifying supply and improving energy security.

Understanding the historical rate of decarbonization—the "background rate"—is central to understanding the problems with the so-called stabilization wedges that were discussed in Chapter 2. Scholars have looked at the historical rate of decarbonization, especially over the period 1980 to 2000, and assumed that it was somehow guaranteed and would thus continue into the future at the same rate as it had during

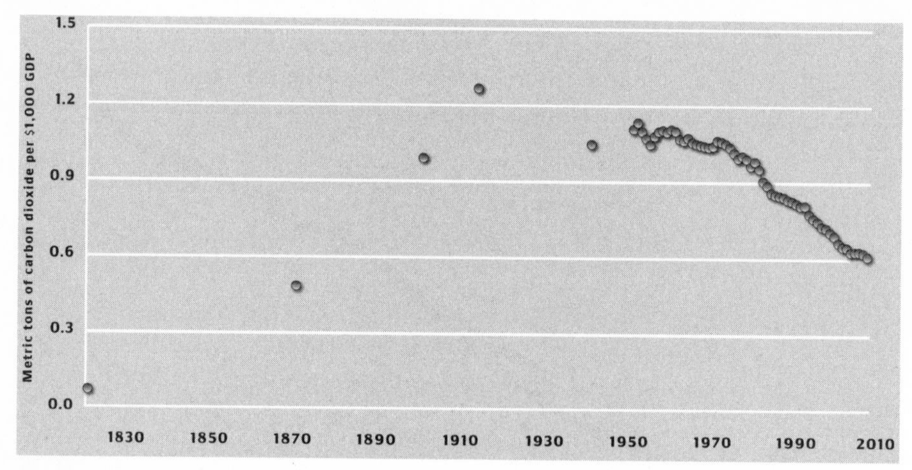

FIGURE 3.4 Carbon dioxide intensity of the global economy, 1820–2006. Source: A. Maddison and U.S. Department of Energy.

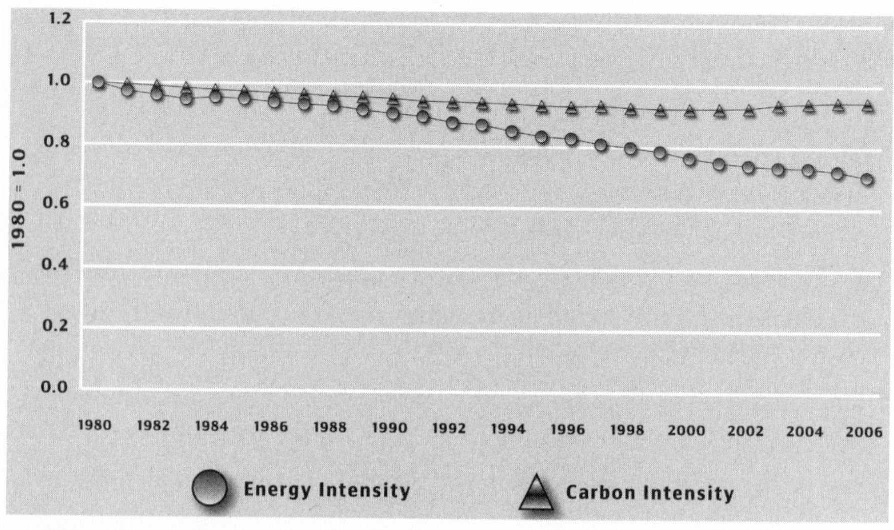

FIGURE 3.5 Energy intensity of GDP and carbon intensity of energy consumption, 1980–2006. Source: U.S. Department of Energy.

those decades (some scholars have even assumed, inexplicably, that the background rate of decarbonization would increase dramatically with no policy actions).[29] Climate policies focused on decarbonization would simply have to piggyback on that background rate, nudging it ahead a bit to meet aggressive stabilization targets.

Unfortunately, there turned out to be two major flaws in that line of thinking. First, we have learned in the first decade of the twenty-first century that "automatic" decarbonization is not so automatic. The second major flaw is that as countries have adopted explicit policies focused on improving energy efficiency and decarbonizing energy supply, it has proven impossible to maintain a clear distinction between "background" decarbonization and that which is the focus of policy, as all policies that have positively influenced the decarbonization of the economy are counted as "climate policies," whether they were actually intended as such or not. Hence, there is a large potential for the double counting (i.e., in assumptions about background rates and a second time as additional efforts in climate policies) of the effects of such policies if they are assumed to be both part of the background and part of explicit climate policies. Double counting is problematic because it can lead us to believe that we are making progress, when we are actually just pursuing policies that approximate business as usual.

Thus, to fully understand the challenge of decarbonization and to avoid the risk of double counting, it is important to start at today's level of carbon dioxide per unit of GDP and then ask what level of decarbonization is implied by a particular stabilization target. Of course, it is not enough to specify the stabilization target; the Kaya Identity tells us that we also have to specify a rate of economic growth as well.

Figure 3.6 shows the implied annual average rates of decarbonization that would be necessary in order for the world to reduce its total carbon dioxide emissions from the burning of fossil fuels to a level 50 percent below 1990 levels by 2050, the level recommended at Copenhagen in December 2009. The figure shows that the global economy would have to decarbonize from 0.62 tonnes of carbon dioxide per $1,000 GDP in 2006 to below 0.20 in 2050 for all rates of GDP growth, and perhaps below 0.10 at higher rates of GDP growth. Global GDP growth was 3.5 percent annually from 1980 to 2006, but the exact future rate is in some respects unimportant, as for annual GDP growth rates of 1 to 5 percent all values of carbon dioxide emissions to GDP are less than 0.20, though the difference between an average 1 percent rate and 5 percent rate is a factor of about 5 in 2050.

Figure 3.7 shows the historical decarbonization of the global economy from 1980 to 2006, as well as the decarbonization curve for 2007

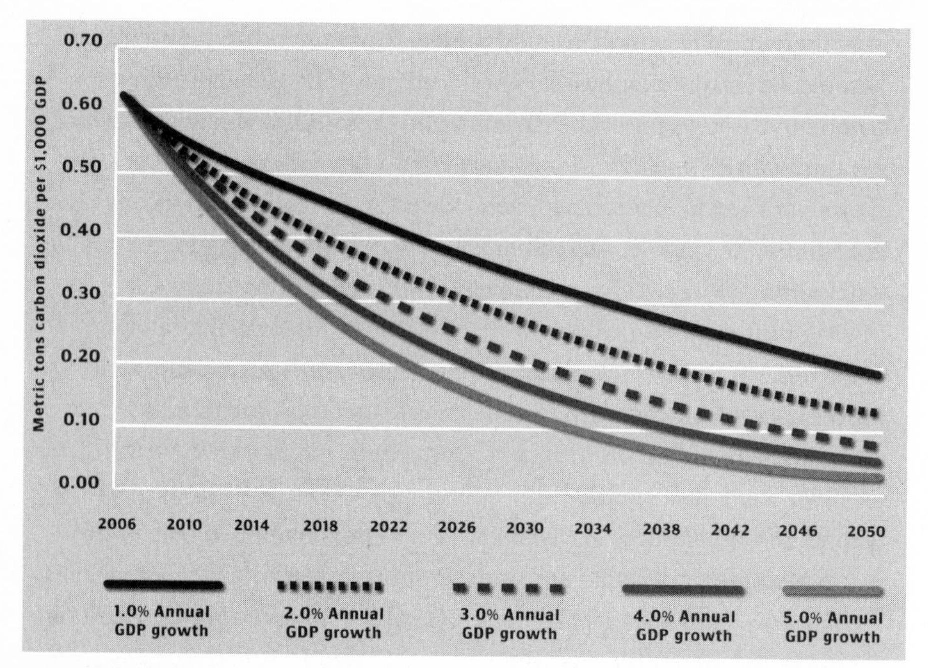

FIGURE 3.6 Implied decarbonization of the global economy. Source: Author's calculations.

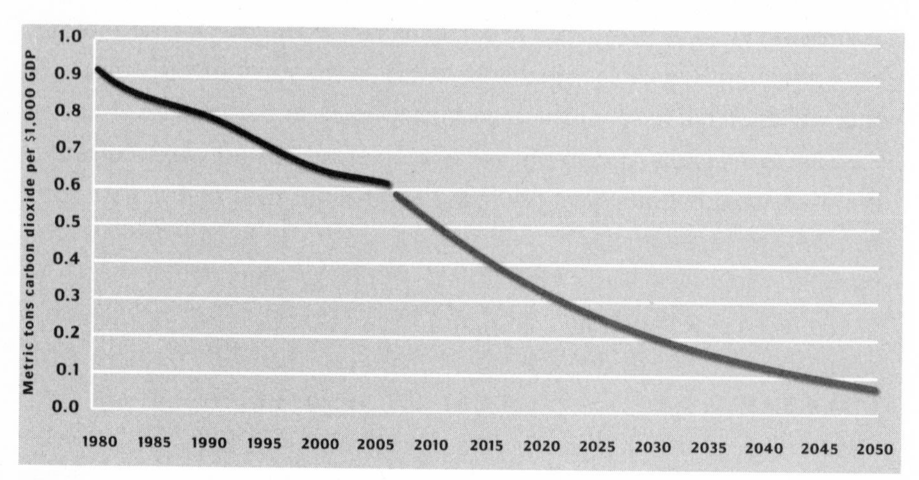

FIGURE 3.7 Historical and projected decarbonization of the global economy. Source: Author's calculations.

to 2050 based on the 3.0 percent (middle) value for future growth from Figure 3.6. It shows that a rapid increase in the average rate of decarbonization would be necessary to achieve a reduction in emissions of 50 percent below 1990 values. An important result from this type of analysis is that the conclusions are qualitatively very much the same for other possible emissions-reduction targets or different assumptions about economic growth. The bottom line is that to stabilize concentrations of carbon dioxide in the atmosphere at low levels will require advances in decarbonization of the global economy beyond that observed over the past decades and even the past century. The average annual rate of decarbonization implied by a 50 percent reduction in emissions below 1990 levels by 2050 for a 3.0 percent annual GDP growth is 4.4 percent per year, whereas the world actually experienced a 1.5 percent rate of decarbonization from 1980 to 2006 while achieving a 3.5 percent average rate of GDP growth.

Is decarbonization to below 0.20 or 0.10 tonnes of carbon dioxide per $1,000 of GDP by 2050 a lot or a little? What does it mean practically? We can better assess what it really means to decarbonize to a particular level by looking at the actual decarbonization experience of countries around the world, and so this is where we go next. Chapter 4 will conclude with a far more intuitive, and sobering, answer to these questions.

CHAPTER 4

Decarbonization Policies Around the World

FIGURE 4.1 SHOWS the share of global emissions of carbon dioxide from the world's top 20 emitters, led by China and the United States, as well as the total emissions of 193 other countries not listed individually.[1] The top 20 emitters were responsible for about 80 percent of total global emissions in 2006.[2] The other 193 countries were responsible for about the same amount as China or the United States. This chapter will survey aspects of domestic decarbonization policies of several of these countries. What we will see is that despite significant effort in many countries, no country has yet figured out how to decarbonize its economy at a pace beyond historical rates, much less the very aggressive rates needed to achieve ambitious emissions-reduction targets. The chapter will then conclude by explaining the significance of the survey for national and international climate policies.

Using the same ratio that was presented at the end of Chapter 3 of metric tons of carbon dioxide emissions from the burning of fossil fuels for each $1,000 of GDP in 2006, Figure 4.2 shows this ratio for each of the top 20 emitting countries.[3] The countries are ordered left to right from highest to lowest emitters (as in Figure 4.1). France, with its large use of nuclear power, has the most carbon-efficient economy of the top 20, and South Africa, with its heavy reliance on coal-powered energy, has the least carbon-efficient economy, emitting more than six times as much carbon dioxide per unit of GDP than does France.

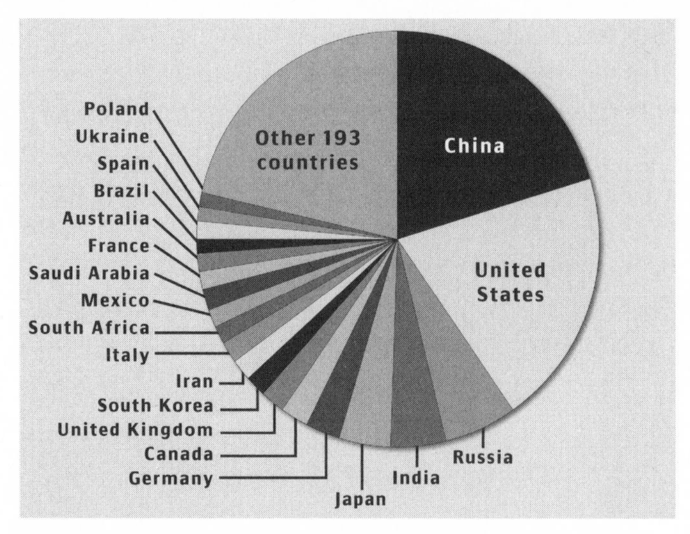

FIGURE 4.1 Global carbon dioxide emissions, 2006. Source: U.S. Department of Energy.

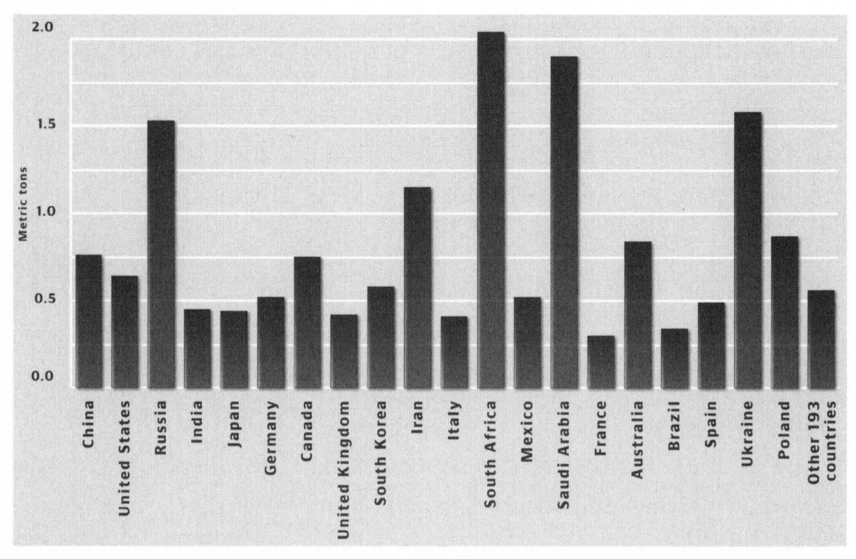

FIGURE 4.2 Metric tons of carbon dioxide per $1,000 2006 GDP. Source: U.S. Department of Energy and A. Maddison.

Let us next look more closely at a few of these countries. In absolute terms Japan, India, and the United Kingdom had relatively more carbon-efficient economies than did Germany, which was somewhat more carbon efficient than the United States, which in turn was more carbon efficient than China and Australia. As we will see, the reasons for

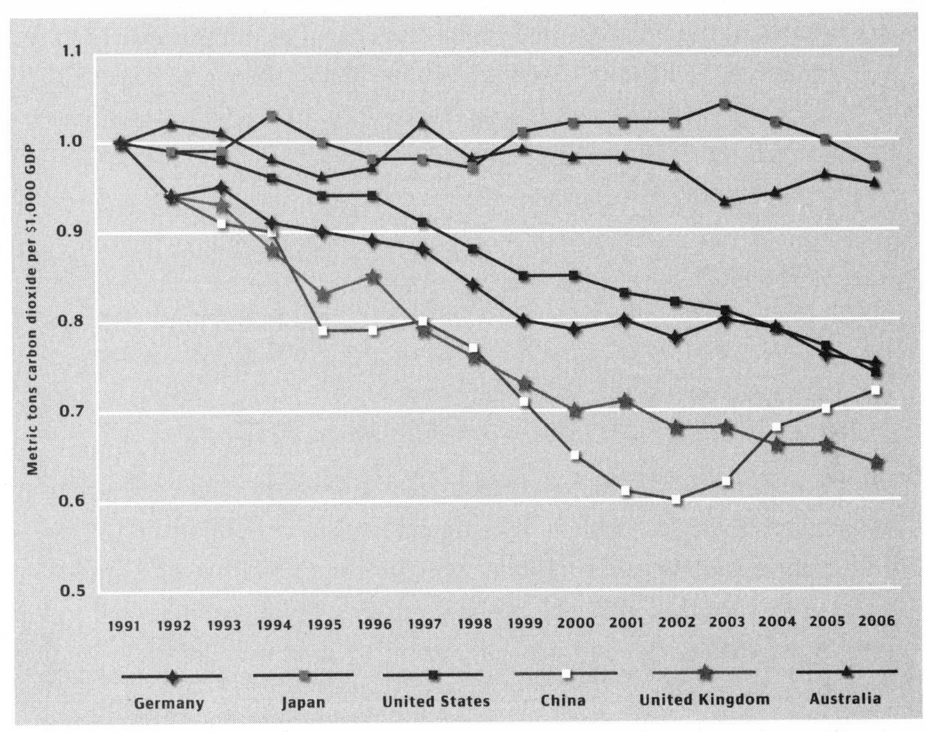

FIGURE 4.3 Relative improvement in CO_2 per GDP, 1991–2006. Source: Author's calculations.

these differences have everything to do with context and history and nothing to do with climate policies.

For each of these six countries, Figure 4.3 shows the relative improvement in carbon dioxide per unit of GDP for the period from 1991 (chosen as the first full year of German reunification) through 2006. The figure shows several interesting trends. First, Japan, even with its relatively low carbon intensity in 2006, has seen little change over a decade and a half, an issue that we will revisit shortly. Germany and the United States start at different absolute levels, but their respective pace of decarbonization was almost identical, despite the fact that the two countries had very different policies and politics during that time period; for instance, Germany signed on to the Kyoto Protocol of the Framework Convention on Climate Change in 1997, whereas the United States rejected it. The U.S. and German experiences indicate that there are many different paths to decarbonization. China experienced a fast rate of decarbonization in the 1990s as its economy grew rapidly due to the effects of globalization. This trend abruptly reversed in the 2000s as China sought to keep pace with an incredible increase in demand for energy, which it met by dramatically expanding its use of carbon-based fuels.[4] Australia saw little change in the carbon intensity of its economy over this period. Let's now look at the policies of several countries in a bit more detail and explore what they signify for future efforts to accelerate decarbonization.

United Kingdom: The Climate Change Act of 2008

On November 26, 2008, the British government enacted the Climate Change Act of 2008, mandating national emissions reductions. In December of that year the United Kingdom's Committee on Climate Change (created by the act) released a report recommending that national greenhouse gas emissions be reduced by at least 80 percent by 2050 and by 34 percent by 2022 (or 42 percent if an international agreement on climate change is reached) from a 1990 baseline. The report argued that this amount of emissions reduction is achievable at an affordable cost of between 1 and 2 percent of GDP in 2050.

In 2006 the UK produced 0.42 metric tons of carbon dioxide for every $1,000 of GDP. Figure 4.4 shows decarbonization in the UK from 1980 to 2006. It also shows the required annual average rates of decarbonization of the UK economy from 2007 to 2050 (for a 2 percent assumed annual GDP growth rate) implied by a target of an 80 percent reduction in carbon dioxide emissions from 1990 levels. The carbon intensity of the UK economy would have to reach a level of 0.02 to 0.05 metric tons of carbon dioxide per $1,000 of GDP by 2050, for faster (3 percent) and slower (1 percent) rates of economic growth respectively. Figure 4.4 also shows the same information for 2022 implied by the target of a 34 percent reduction in carbon dioxide levels from 1990. The carbon intensity of the UK economy would have to reach a level of 0.17 (for 3 percent annual GDP growth) to 0.24 (for 1 percent annual GDP growth) metric tons of carbon dioxide per $1,000 of GDP by 2022, from 0.42 in 2006.

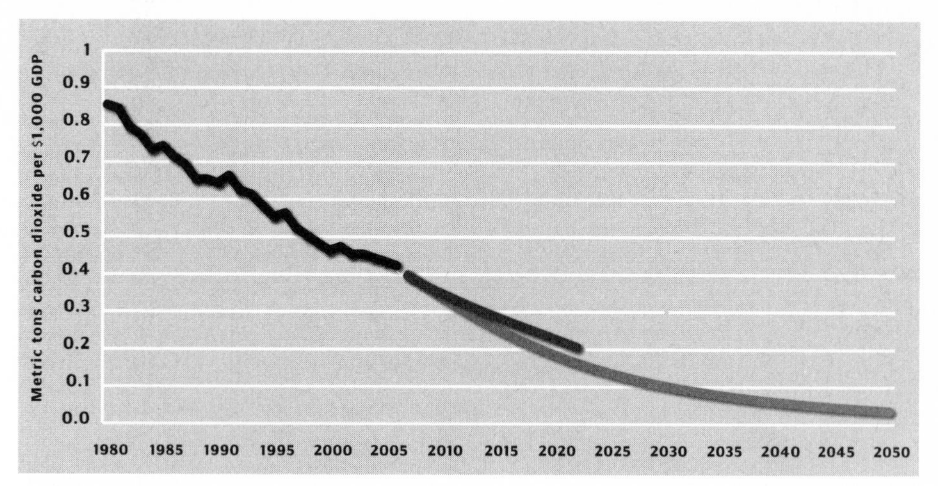

FIGURE 4.4 Historical and implied decarbonization of the UK economy. Source: Author's calculations.

The implied rates of decarbonization of the UK economy for the curves in Figure 4.4 are 4.4 percent per year for the 2022 target and 5.5 percent for the 2050 target. These numbers are substantially higher than the rates of decarbonization observed from 1980 to 2006 and 2001 to 2006, as summarized in Table 4.1.

Achieving the ambitious targets for emissions reductions set forth in the UK Climate Change Act will require rates of decarbonization much higher than have been achieved in any major economy in recent decades. The Climate Change Committee has not addressed explicitly whether this is a reasonable goal. However, in an interview, Julia King, vice chancellor of Aston University in Birmingham and member of the Climate Change Committee, responded to an earlier version of this analysis by saying that in fact the scenarios provided by the committee have "been tested for do-ability." King apparently meant theoretical technical "do-ability" (along the lines discussed in Chapter 2 in the section "Do We Have All the Technology We Need?"), as she also explained that achieving the targets has both technical and political challenges, with the latter difficult to overcome: "I think you really do need to take due account of the fact that most people who are putting together targets and timetables are doing this on the basis of a lot of research into potential scenarios. It's another issue turning that into policy, for governments, and it's

TABLE 4.1 Annual rate of decarbonization of the UK economy observed (first two columns) for 1980 to 2006 and 2001 to 2006, and implied by the 2022 and 2050 targets assuming 2.0 percent future GDP growth

	1980–2006	*2001–2006*	*2007–2022*	*2007–2050*
Actual	–1.9 percent	–1.3 percent		
Implied by targets			–4.4 percent	–5.5 percent

very easy for all of us who don't have to be elected to say, 'This is how I would do it,' and I have a lot of sympathy for our politicians, because they are dealing with extremely selfish populations."[5]

A key aspect to effective policy implementation is that policies must be not just technically feasible but also socially and politically acceptable. For instance, it is one thing to say that deployment of, say, dozens of nuclear power plants is technically possible; it is quite another to achieve it in practice. Regardless of the technical arguments for theoretical "do-ability," the targets of the Climate Change Act fail the test of practical "do-ability," as we will see.

One important reason for the decarbonization of the UK's economy is that manufacturing has declined as a portion of its economy, from 33 percent in 1970 to 13 percent in 2007.[6] A onetime switch from coal to gas—the so-called dash for gas motivated by Margaret Thatcher's policies with respect to unions and state control of energy—also played a role. Reliance on such actions in the future is obviously not a sustainable route to decarbonization. Further, there is no recent precedent among developed countries with large economies for the sustained rapid rates of decarbonization implied by the Climate Change Act. Such rates necessarily must be several times greater than observed in the UK in recent decades.

France, which of the major economies has the lowest ratio of emissions to GDP, provides a good point of comparison for the UK. France has achieved its relatively low level of decarbonization due to its reliance on nuclear power for electricity generation. France achieved an average rate of decarbonization of about 2.5 percent per year from 1980 to 2006, but achieved only about 1.0 percent per year from 1990 to

2006. It took France about twenty years to decarbonize from 0.42 metric tons of carbon dioxide per $1,000 GDP, the level of the UK in 2006, to 0.30 metric tons of carbon dioxide per $1,000 GDP.

In order for the United Kingdom to achieve the very low ratios of carbon emissions to GDP implied by its policy targets, it must at some point reach France's ratio of 0.30 along the way. For the UK to be on pace to achieve the targets for emissions reductions implied by the Climate Change Act, its economy would have to become as carbon efficient as France's by no later than 2015 (depending on economic growth). See Figure 4.4 above and, in particular, the year in which the implied decarbonization curve crosses 0.30. In practical terms this level of decarbonization of the UK economy could be achieved, for example, with a level of effort equivalent to building and operating about forty new nuclear power stations by 2015, displacing coal- and gas-fired electrical generation.[7] An example of the sort of nuclear power station used in this analysis is the Dungeness B station in Kent, on the southeast coast of England.[8]

Following that achievement, to meet the 2022 target the UK would then have to decarbonize by an additional 33 percent, that is, from 0.30 metric tons of carbon dioxide per $1,000 GDP, to 0.20 metric tons. The analysis of the Climate Change Committee is largely consistent with this conclusion, explaining that achievement of the 2050 target would require that all UK electricity generation be completely decarbonized by 2030.[9]

Upon reading an early draft of this analysis, Colin Challen, member of Parliament from the Labor Party and chairman of its All Party Parliamentary Climate Change Group, commented to the BBC that he agreed with the analysis, making reference to the government's recent decision to expand Heathrow Airport despite the fact that an expansion would lead to increased emissions from greater air travel:[10]

> This [analysis] raises questions which I do not think have been factored into the thinking behind the Climate Change Act. The task [of cutting emissions by 80 percent from 1990 levels by 2050] is already staggeringly huge and, as we have seen, well beyond our current

> political capacity to deliver. Heathrow is a prime example of ducking the responsibility. It is hard to see any tough choices being made in the current climate. A greater population implies more embedded carbon-dioxide emissions in imported goods, but the climate change committee is only empowered to consider domestic emissions.[11]

Given the magnitude of the challenge and the pace of action, it would not be too strong a conclusion to suggest that the UK Climate Change Act has failed even before it has gotten started. The Climate Change Act does have a provision for the relevant government official to amend the targets and timetable, but apparently not in the case of a failure to meet the targets. It seems likely that the Climate Change Act will have to be revisited by Parliament or simply ignored by policy makers. Achievement of its targets is simply and obviously not a realistic option.

Japan: A Genuine Clear-Water Climate Policy?

In June 2009 Japanese prime minister Taro Aso announced that Japan would seek to reduce its greenhouse gas emissions by 15 percent from 2005 levels by 2020. The prime minister said, "The target we are using is for 'genuine clear water' or *mamizu* as we say in Japanese—truly a genuine net effect of our effort to save and conserve energy."[12] The word *mamizu* is often used in Japanese politics when discussing the difference between, for example, actual budget cuts and those that might simply be tricks of accounting. In the context of climate policy a *mamizu* climate policy refers to purely domestic efforts, not counting on emission reductions accounted for using carbon offsets or land-use changes. It thus refers to explicit efforts to accelerate decarbonization of the Japanese economy.

Immediately upon announcing its proposed *mamizu* targets, the Japanese government was harshly criticized for its lack of commitment and vision. Yvo de Boer, director of the United Nations Framework Convention on Climate Change, commented that the commitment fell far short of what was needed, saying that the Japanese proposal left him speechless.[13] Facing a barrage of criticism, several weeks later Japan appeared to soften its stance on its *mamizu* climate policy when environment minister Tetsuo Saito announced that Japan would be willing to consider adding to its target by using international mechanisms such as offsets.[14] Further change in targets came in August 2009, when the Democratic Party of Japan (DPJ) unseated the Liberal Democratic Party (LDP), which had held power almost exclusively since 1955. The change in government was accompanied by a major change in Japan's proposed target for emissions reductions, which was dramatically increased to a 25 percent reduction from 1990 levels by 2020, equivalent to about a 37 percent decrease from 2005 values. Were either the LDP or DPJ emissions-reduction targets reasonable? Like the UK case, the analysis is as simple as it is sobering.

Figure 4.5 shows the actual rate of decarbonization of the Japanese economy from 1980 to 2006 as well as the rates of decarbonization implied by the 2020 and 2050 targets assuming an average 1.5 percent annual GDP growth. In 2006 Japan produced 0.42 metric tons of carbon dioxide for every $1,000 of GDP. To achieve an 80 percent reduction in its emissions from 1990 levels by 2050 implies that the carbon intensity of the Japanese economy would have to reach a level of 0.02 to 0.06 metric tons of carbon dioxide per $1,000 of GDP (for average annual GDP growth rates of 3 percent and 1 percent, respectively). Figure 4.5 also shows the decarbonization to 2020 implied by the *mamizu* (LDP) target of a 15 percent reduction in carbon dioxide levels from 2005 and the DPJ target of a 25 percent reduction below 1990 levels (for a 1.5 percent average GDP growth rate). The figure shows that the carbon intensity of the Japanese economy would have to reach a level of decarbonization about equal to that of France (in 2006) by 2014 or 2020, for the respective targets.

The rates of decarbonization of the Japanese economy implied by the targets can be seen in Table 4.2. These numbers are substantially higher than the rates of decarbonization observed from 1980 to 2006 and 2001 to 2006. Japan faced a range of criticism when it announced its 2020 *mamizu* target to reduce its domestic emissions by 15 percent from 2005 levels by 2020. Based on this analysis above, such criticism was unfounded for several reasons.

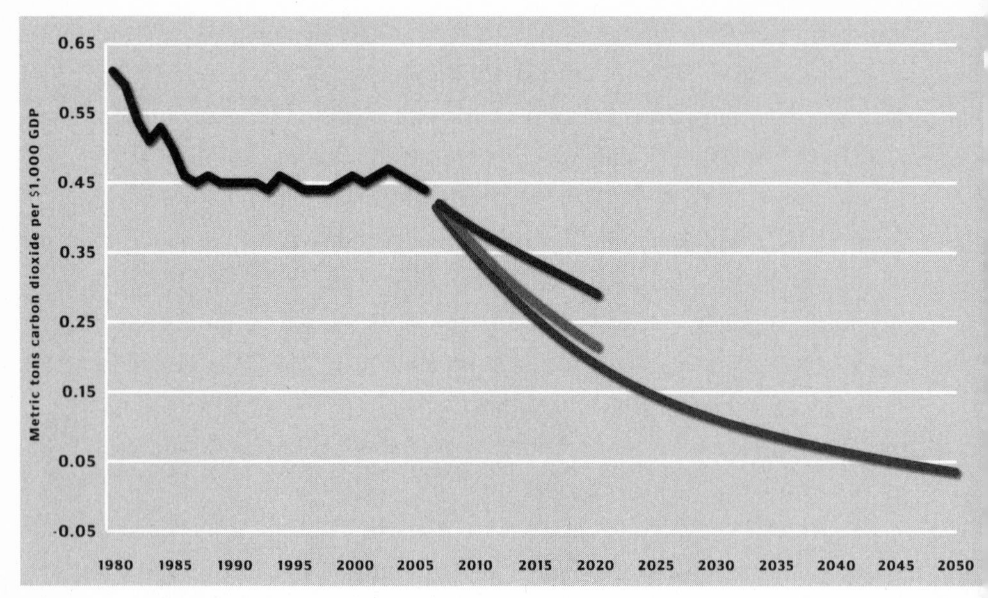

FIGURE 4.5 Historical and projected decarbonization of Japan's economy. Source: Author's calculations.

TABLE 4.2 Annual rate of decarbonization of the Japanese economy observed (first two columns) for 1980 to 2006 and 2001 to 2006, and implied (third and fourth columns) by the 2020 and 2050 targets of 15 percent of 2005 levels, 25 percent below 1990, and 80 percent below 1990, respectively (assuming 1.5 percent future GDP growth)

	1980–2006	*2001–2006*	*2007–2022*	*2007–2050*
Actual	–1.3 percent	–0.9 percent		
Implied by 15 percent reduction below 2005 by 2020			–2.6 percent	
Implied by 25 percent reduction below 1990 by 2020			–4.6 percent	
Implied by 80 percent reduction below 1990 by 2050				–5.4 percent

First, the rate of decarbonization implied by the 2020 target is twice its historical rate, implying substantial effort. Because no one knows how fast a major economy can decarbonize, there seems little point in arguing about proposed rates of decarbonization well outside that which actually has been possible. Policy implementation will be the ultimate arbiter of such proposals. There is essentially no qualitative difference between the Japanese and UK decarbonization targets, as in both instances the various targets imply a rate of decarbonization far outside the range of each country's experience for periods of a decade or longer. Both countries' targets appear unlikely to be met, though arguably the *mamizu* policy is more realistic than the UK Climate Change Act or the aggressive DPJ target.

Second, the rate of decarbonization in the Japanese 2020 targets is in excess of that which has been observed in any major economy in recent decades. However, Japan's experience during the early 1980s provides a notable exception: from 1980 to 1986 the average decarbonization of the Japanese economy was 4.4 percent per year. Vaclav Smil of the University of Manitoba argues that this achievement was due to a preponderance of "low hanging fruit" and is unlikely to be replicated, much less sustained, in the future (see Figure 4.3).[15] The shift in the Japanese economy from carbon-intensive industries, especially aluminum production, to less carbon-intensive industries also played an important role. Today, Japan is already one of the most carbon-efficient economies in the world, thereby making further gains more difficult and expensive than they would be in the generally less efficient economies of North America and Europe. Japan may be an important test case in the limits to efficiency gains as a strategy of decarbonization.[16] Thus, a *mamizu* approach to climate policy would provide valuable experience on how fast decarbonization rates might be accelerated.

An analysis by Professor Tetsuo Yuhara of the University of Tokyo explained the steps that Japan would need to take to meet the 25 percent reduction target below 1990 levels by 2020:[17]

1. Solar power generation must increase by 55 percent from current levels requiring photovoltaic cells to be installed in all new

houses and some existing houses (for a total of 600,000 installations annually).

2. Fifteen new nuclear power plants must be built and operated with 90 percent capacity rate (far above the current rate of 60 percent).
3. Increased thermal power from both gas power plants and biomass mixed combustion would be needed.
4. Ninety percent of sales of new vehicles must be of next generation vehicles (i.e., hybrid or electric cars).
5. All new houses and existing houses must have heat insulation installed, and mandatory energy conservation standards must be implemented.
6. The price of one ton of carbon dioxide would be 82,000 yen (~$80), compared to 15,000 yen (~$15) for the previous target of an 8 percent reduction, or the current price of around 7,000 yen (~$7).

These are undoubtedly ambitious (some might say impossible) goals. For instance, the proposal to deploy fifteen new nuclear power plants within a decade appears to stretch the bounds of credulity, even though Japan does have the third-most nuclear plants in the world (after the United States and France) and has plans to build more.[18] Japan's adoption of aggressive but impossible-to-achieve targets for emissions reductions signifies a desire to meet the symbolic needs of international climate politics while sacrificing the practical challenge of decarbonization policy. If Japan's *mamizu* targets were to be criticized, it should have been because they were too aggressive, not because they were too weak.

Australia: The Ups and Downs of an Emissions Trading Scheme

On December 12, 2007, Australian prime minister Kevin Rudd, having been sworn into office only the week before, gave a rousing speech at the Thirteenth Conference of the Parties to the United Nations Framework Convention on Climate Change, held in Bali, Indonesia.

Rudd explained that Australia was ready to commit to binding targets for emissions reductions. He promised to cut greenhouse emissions by 60 percent from 2000 levels by 2050. He had commissioned a study, known as the Garnaut Review, which was due in mid-2008. He insisted: "These will be real targets. These will be robust targets. And they will be targets fully cognizant of the science. . . . But it is not enough just to have targets. We have to be prepared to back them with sustained action—because targets must be, must be translated into reality. Australia will implement a comprehensive emissions trading scheme by 2010 to deliver these targets."[19] At Bali, Prime Minister Rudd was met with "long and loud applause."[20] However, despite signing the Kyoto Protocol as his first official act as prime minister and delivering the rousing Bali speech, Australia soon found itself facing international criticism for its failure to announce any short-term targets at the Bali meeting.[21]

Upon the release of an interim draft of the Garnaut Review in February 2008, its author, Ross Garnaut of the Australian National University, called for Australia to increase its targets beyond those mentioned at Bali: "Australia should be ready to go beyond its stated 60 percent reduction target by 2050 in an effective global agreement that includes developing nations."[22] Immediately thereafter, Prime Minster Rudd's government appeared to distance itself from the report. Climate Change Minister Penny Wong said of the report's conclusions, "We welcome Professor Garnaut's input. . . . [O]f course we will also be looking at other inputs, such as modelling from the Australian Treasury," prompting the leader of the Australian Green Party to complain that "Penny Wong has reduced Ross Garnaut to 'input.'"[23]

Less than two weeks after the draft Garnaut Review was released the Rudd government released a "green paper" outlining its initial plans for a Carbon Pollution Reduction Scheme, a policy based on a cap-and-trade approach to emissions reductions along the lines of the European Emissions Trading Scheme (ETS). A white paper outlining the final plans for the proposed CPRS was subsequently released in December 2008, as the Australian government announced an emissions-reduction target of between 5 percent (unilaterally) and 15 percent (in concert with other nations) below 2020 levels, and a proposed 60 percent reduction by 2050.[24] In the face of severe criticism for its lack of ambition,[25] the government justified its target in terms of the implications for per capita emissions, which it argued were on par with those promised by other nations. Before long, however, the Rudd government responded to its critics by raising its targets: in May 2009 the interim target was increased to a 25 percent reduction even as the proposed starting implementation date for the CPRS was delayed to 2011, justified on the need to allow the economy to regain strength in the aftermath of the global financial crisis.[26]

Regardless, in August 2009 the Australian Senate voted down the CPRS, prompting the government to split the renewable-energy provisions from the trading-scheme provisions. The renewable-energy package was subsequently passed into law. In November 2009 the opposition Liberal Party saw a revolt over the proposed CPRS, resulting in a change in party leadership and a second defeat for the trading scheme in the Senate. In the spring of 2010 the Australian carbon-trading scheme was delayed again, and the opposition party used the issue to gain support among the Australian populace. While debate over the ETS continues and its legislative future is uncertain, it is not too early to conduct an assessment of the various targets implied by the ETS. How realistic are Australia's proposed emissions-reduction goals in the short and long terms?

In 2006 Australia produced 0.84 metric tons of carbon dioxide for every $1,000 of GDP. Figure 4.6 shows the actual annual rate of decarbonization of the Australian economy from 1980 to 2006. The figure also shows the implied decarbonization for emissions-reduction targets of 5 percent, 15 percent, and 25 percent from 2000 levels by 2020 as well as for a 60 percent reduction by 2050, for an annual average 2.5 percent GDP growth rate.[27] The figure shows that the carbon intensity of the Australian economy would have to reach a level of about 0.10 metric tons of carbon dioxide per $1,000 of GDP by 2050, from 0.84 in 2006. The figure shows that the carbon intensity of the Australian economy would have to be cut by about a third to more

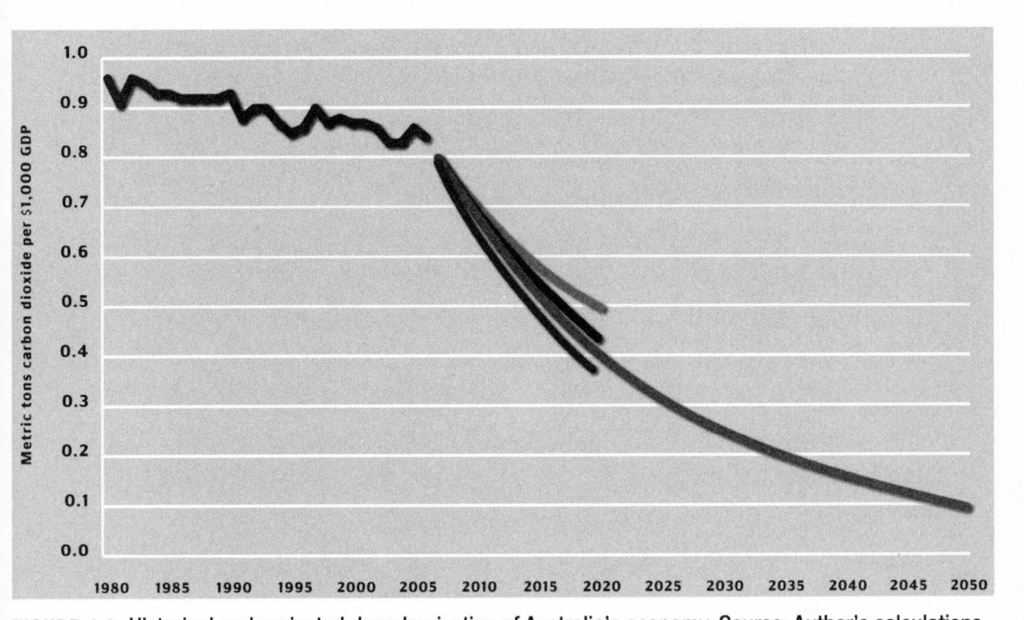

FIGURE 4.6 Historical and projected decarbonization of Australia's economy. Source: Author's calculations.

than half by 2020, depending upon assumptions, from its value of 0.84 in 2006.

The targets imply that Australia would have to achieve the 2006 emissions intensity of Japan by no later than 2018 for a 25 percent reduction target, by 2020 for a 15 percent reduction target, or by 2023 for a 5 percent reduction target (see Table 4.3). Japan has a highly efficient economy on several small islands with almost no domestic energy resources and operates a sizable number of nuclear power plants. Australia, on the other hand, burns much more coal and is generally profligate with carbon.

To think that Australia could achieve Japanese levels of decarbonization within the next decade strains credulity. This view was reinforced by Australia's climate change minister, Penny Wong, who commented on an earlier version of this analysis in February 2010, explaining that it neglected "the important role international permits will play in Australia's low cost transition to a low pollution future."[28] By "international permits" she was referring to carbon offsets, which are discussed in some depth later in the chapter. For now, what is important to

TABLE 4.3 Annual rate of decarbonization of the Australian economy observed (first two columns) for 1980 to 2006 and 2001 to 2006, and implied (third and fourth columns) by the 2020 and 2050 targets

	1980–2006	*2001–2006*	*2007–2020*	*2007–2050*
Actual	–0.5 percent	–0.7 percent		
Implied (5 percent reduction target, 2.5 percent GDP growth)			–3.8 percent	
Implied (15 percent reduction target, 2.5 percent GDP growth)			–4.6 percent	
Implied (25 percent reduction target, 2.5 percent GDP growth)			–5.4 percent	
Implied (60 percent reduction target, 2.5 percent GDP growth)				–4.8 percent

understand is that the use of "international permits" implies limited changes in the decarbonization of the Australian economy. They would not be, as the Japanese say, "genuine clear water."

Another way to look at the magnitude of the challenge of decarbonizing the Australian economy is in terms of its energy mix. It is straightforward to convert the energy mix into greenhouse gas emissions by multiplying the amount of energy consumed in quads by the amount of carbon emitted per quad for each fuel.[29] According to the U.S. Energy Information Agency, in 2004 Australia emitted about 391 million metric tons (Mt) of carbon dioxide from 5.3 quads of consumption, with the mix shown in Figure 4.7.[30] Multiplying the carbon dioxide generated per quad (shown in Figure 4.8) by the proportion of energy from each fuel source results in 390 Mt of carbon, essentially the same as that reported by the U.S. EIA.

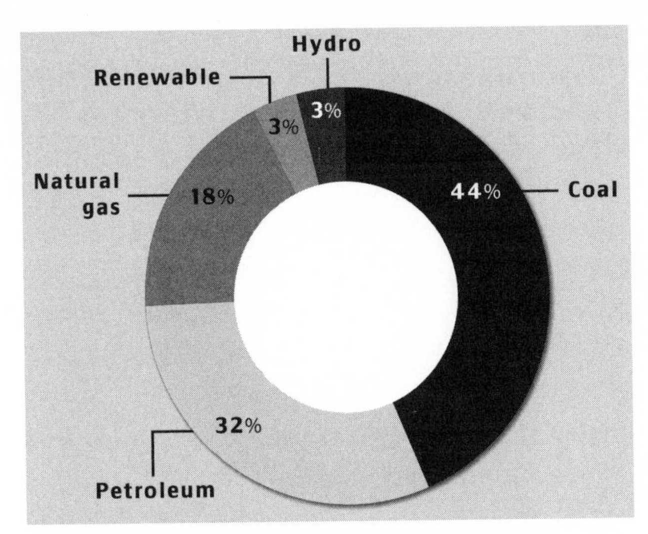

FIGURE 4.7 Australia's energy mix. Source: U.S. Energy Information Administration.

With this information it is then possible to perform a simple sensitivity analysis describing what it would take to decarbonize the Australian economy to a level consistent with a particular emissions-reduction target. In 2004 Australia produced 0.83 metric tons of carbon dioxide emissions per $1,000 (U.S.) (essentially the same as in 2006). To cut this amount in half over the next decade or less—as implied by the 5 percent, 15 percent, and 25 percent 2020 targets—would require that nearly all Australian coal consumption be replaced by a zero-carbon alternative such as nuclear or renewable energy. If an average nuclear plant provides 750 megawatts of electricity[31] and one quad is equivalent to 11,000 megawatts of electricity,[32] then about fifteen nuclear power plants would provide 1 quad. Coal provided 2.4 quads for Australia in 2004, meaning that it could be replaced by about thirty-six nuclear power plants.

Of course, Australia's energy consumption has increased since 2004 and is expected to increase in the future. If Australia's demand for energy increases by 1.5 percent per year to 2020, then an additional 1.4 quads of energy will be needed, implying the equivalent of twenty-one additional nuclear power plants, for a total of fifty-seven. These assumptions can be adjusted to explore the implications of aggressive

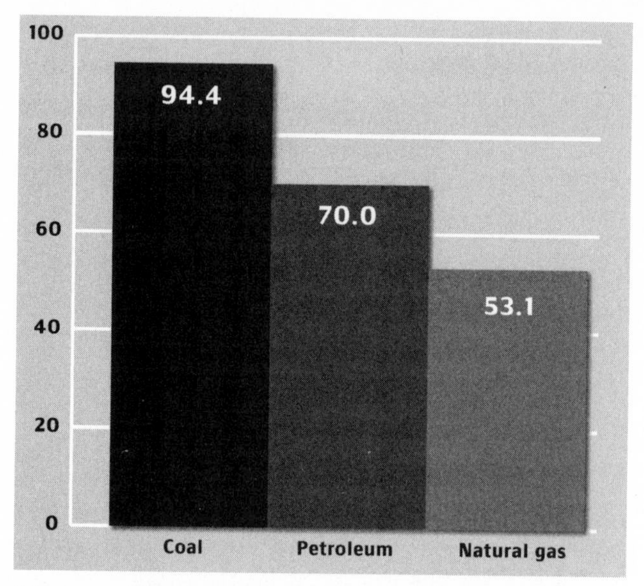

FIGURE 4.8 Million metric tons of carbon dioxide per quad of energy. Source: Calculated from data provided by the U.S. Energy Information Administration.

energy-efficiency programs or expansion of renewable-energy technologies (or other assumptions, such as the expansion of natural gas). For instance, if demand is held constant at 2004 levels and renewable energy constitutes 20 percent of the total 2004 energy mix, then only thirteen 750-megawatt nuclear power plants would be needed by 2020. Different assumptions will, of course, lead to different results, and the ones presented above are intended to be illustrative of the magnitude of the decarbonization challenge under a reasonable set of assumptions. The conclusion that the magnitude of the challenge is enormous is not particularly sensitive to these assumptions. If Australia relies on "international permits" to meet its emissions-reduction targets, as implied by its climate change minister, it would have to use the permits for the majority of the task, under any of the scenarios.

Several Australian readers of an early version of this analysis commented that a comparison of nuclear power plant equivalents, even if hypothetical, would not make much sense to many readers, because Australia has a long history of opposition to nuclear power plants—even building one plant would be an enormous achievement. The same sort of hypothetical sensitivity analysis can be conducted with technologies based on existing solar power plants. The Cloncurry solar thermal power plant in Queensland provides 10 megawatts of electricity.[33] If it operates at 33 percent efficiency, 1 quad of energy could be provided by 3,333 Cloncurry plants. Providing 3.8 quads implies 12,665 such plants, or about 24 plants coming online each week from 2010 to 2020.

What this sensitivity analysis clearly indicates is that to meet proposed emissions-reduction targets, Australia would need to undertake a herculean effort. The level of effort is daunting no matter what sort of technologies are used to illustrate the magnitude of the challenge, even if coupled with very aggressive efforts to increase efficiency and the use of renewable-energy sources. The use of offsets, as we will see, is an example of a sort of "magical thinking" that tends to show up in the climate debate rather than confront the real challenges of decarbonization. Regardless of the nature of the legislation ultimately adopted in Australia, the actual decarbonization of the Australian economy will all but certainly fall short of the proposed targets.

United States: Rejoining the Global Community

After years of U.S. disengagement from international negotiations under the Climate Convention during the presidency of George W. Bush, the Obama administration came into office in 2009 promising a renewed emphasis on climate policy. Subsequently, President Obama proposed a 14 percent reduction in 2020 emissions from a 2005 baseline, and legislation passed subsequently by the House of Representatives in the summer of 2009 mandated a 17 percent reduction. With the Senate not acting on climate policy in 2009, the United States proposed a 17 percent reduction in its carbon dioxide emissions (from a 2005 baseline) by 2020 at the UN climate negotiations in Copenhagen in December 2009. As of this writing the Senate has yet to pass any legislation in support of that goal, and all indications are that if any climate legislation is passed in 2010, it will not include provisions for a so-called cap-and-trade program. But whether it does or does not, the outcome with respect to emissions is all but certain to be much the same.

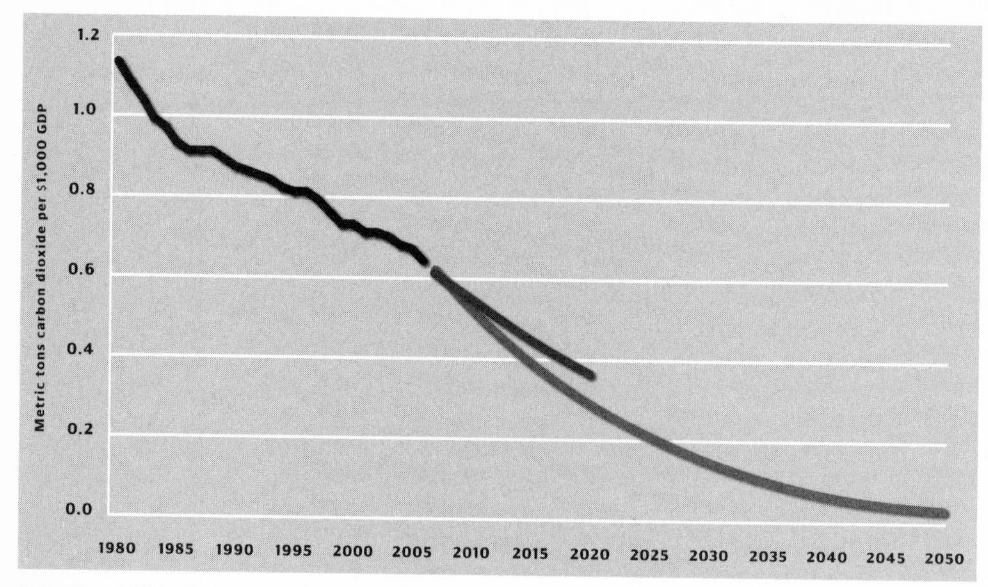

FIGURE 4.9 Historical and implied decarbonization of the U.S. economy. Source: Author's calculations.

By now the analysis is familiar. Like the United Kingdom, Japan, and Australia, the emissions-reduction target proposed by the United States at Copenhagen implies a massive level of effort. That level of effort is insensitive to a target that is a few percent larger or smaller. Figure 4.9 shows the decarbonization implied by a 17 percent reduction (from 2005) by 2020 and an 80 percent reduction (from 2005) by 2050.

The 17 percent reduction target implies achieving the carbon intensity of France by 2026, while the 2050 target implies achieving that level by 2019. What would such an achievement imply in practical terms?

Just as with Australia, the mathematics of U.S. carbon dioxide emissions are not complicated. In 2006 the United States consumed a total of 99.2 quads of energy. Achieving the carbon intensity of France would require that about 57 percent of 2006 coal energy (22.5 quads) be replaced by a carbon-free alternative. If these 12.9 quads were replaced by nuclear power (assuming a 750-MW nuclear plant, as above), this would imply a need for the equivalent of 189 new nuclear power stations. Because energy demand is expected to increase, new demand would also have to be met with a carbon-free alternative. Assuming a

0.5 percent annual increase in energy consumption implies a need for 10.4 new quads above 2006 values, or about an additional 153 nuclear power plants, for a grand total of 342! That seems rather unlikely.

We can perform other sorts of thought experiments with the simple math of emissions and decarbonization. In fact, this is exactly what climate policy experts do in a more complex and precise manner in the form of energy scenarios. To make emissions-reduction math work out in desired ways requires introducing a wide range of assumptions. However, making the scenarios analyzed more complex does not make meeting the challenge in the real world any easier than implied via the simple analysis presented here.

Consider a few examples of scenario analysis with the goal of reaching a 17 percent reduction target below 2005 carbon dioxide emissions in 2020:

Natural gas. Natural gas has been much discussed because it generates less carbon dioxide emissions than does coal for a given amount of energy. However, natural gas is not a long-term solution if the goal of mitigation policy is ultimately a reduction in emission of 80 percent or more. Consider a hypothetical case in which all present and future U.S. coal use is replaced by natural gas to 2020. Carbon dioxide emissions would be only 16 percent less than a 2005 baseline. Unless it is associated with some form of carbon capture and storage, using natural gas to pursue short-term goals would scuttle meeting long-term ones.

Very low carbon-energy sources. For wind and solar to displace enough coal to reach the 17 percent target by 2020 would require that they increase by a factor of twenty-five in absolute terms from their 2008 production of 0.61 quads.[34] Such an increase implies a need for about 200,000 2.5-MW wind turbines of the sort being deployed in West Texas as part of a 600-MW wind farm initiated in 2009 (and this analysis ignores nontrivial issues of intermittency of supply and energy storage and transport).[35] President Obama has expressed a goal of tripling wind- and solar-energy supply during his presidency.

Efficiency. Although there is undoubtedly potential to increase energy efficiency, to reach the 17 percent reduction from 2005 emissions would require a reduction of U.S. energy use by about 2 quads per year for the next decade, equivalent to the shutting down of about 20 power plants per year, ultimately reaching levels of energy consumption last seen in the late 1980s.[36] Assuming that policy makers and citizens want economic growth to continue, this would be a herculean task. With most estimates of future energy demand already assuming significant improvements in efficiency, the task could be even larger if these assumed gains do not occur or if economic growth happens at a faster rate than assumed.

In reality, of course, none of these idealized examples would be applied alone; accelerated decarbonization will require a combination of approaches. However, it is difficult to envisage a scenario that achieves the proposed reductions on the timescale implied by the targets. Achieving the equivalent of deploying more than 300 nuclear power plants in a decade is an enormous task no matter how the scenarios are put together.

Based on the data and analysis in Chapters 3 and 4, you are now empowered to do emissions-reduction math for yourself. Can you see a realistic way for the United States (or the UK, Japan, or Australia) to meet emissions reductions targets with existing technologies?

China, India, Europe, the Others in the Global Top 20, and the Bottom 193

The four countries examined in detail so far represent just under 30 percent of the total (as of 2006) global carbon dioxide emissions. What about the rest of the world?

No discussion of carbon dioxide emissions would be complete without discussing China and India, which were responsible for about 21 percent and 4.4 percent of 2006 emissions, respectively. Both countries are projected to be responsible for an increasing share of global emissions as their economies continue rapid growth. But from those countries' perspective—or indeed, from the perspectives of Brazil, Mexico,

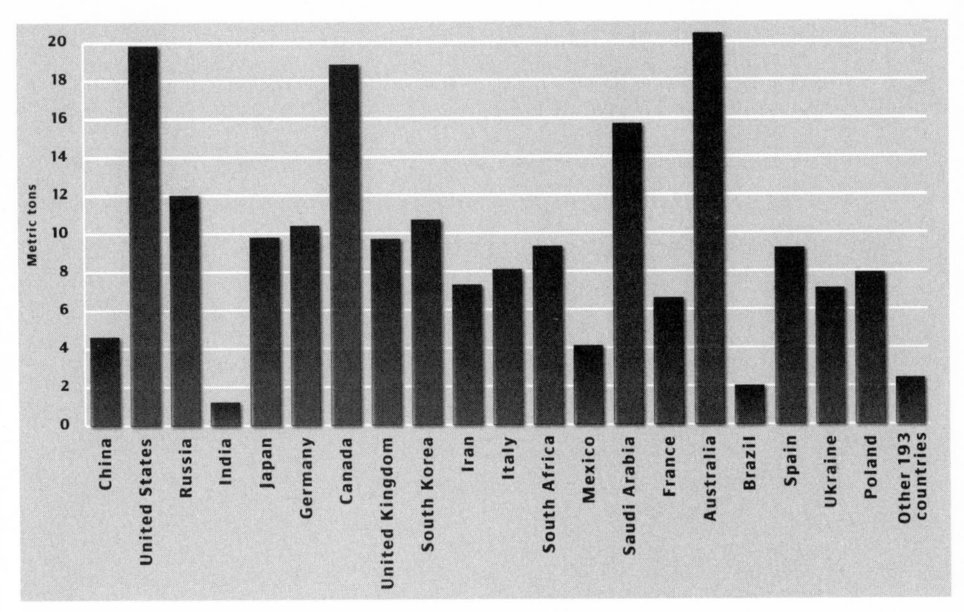

FIGURE 4.10 Per capita carbon dioxide emissions, 2006. Source: Author's calculations.

and the 193 countries outside the top 20 emitters—the mathematics of emissions look very different from a simple tabulation of total national emissions. Figure 4.10 shows the 2006 per capita emissions by country, with the largest total emitter on the left and the smallest on the right. The figure shows a marked difference in per capita emissions among countries often (and in some cases perhaps misleadingly) labeled as "developing" and those that are labeled as "developed." Consider that if China and India had per capita emissions in 2006 equal to that of France, with the lowest per capita emissions among developed countries, global emissions would have been about 30 percent higher. Putting Brazil, Mexico, and the 193 other countries at 2006 French per capita levels would add another 42 percent to 2006 levels.

Because economic growth is tightly coupled to emissions, as we saw in Chapter 3, many of the developing countries have been adamant that taking on emissions-reduction goals is simply not in the cards.[37] India has been particularly explicit about the primacy of economic growth. In summer 2009, Jairam Ramesh, the India environment minister, made this point without nuance: "India will not accept any emission-reduction target—period. This is a non-negotiable stand." The Indian prime minister told a domestic audience, "There is a lot of pressure on India and China on the issue of climate change. We have to resist it." Rajendra Pachauri, head of the IPCC and also an Indian, explained the underlying logic: "Obviously you are not going to ask a country that has 400 million people without a light bulb in their homes to do the same as a country that has splurge of energy."[38]

China has been much more circumspect than India in its statements about emissions reductions, but no less focused on the importance of economic growth. In 2009, for the first time, China's consumers bought more automobiles than consumers in the United States. China also saw more sales of desktop computers than did the United States, and the average size of newly purchased flat-panel TVs was larger.[39] China has explained that its "one-child" policy represents its contributions to international climate policy. Zhao Baige, vice minister of China's National Population and Family Planning Commission, asserted in late 2009 that China's one-child policy had prevented 400 million births that otherwise would have occurred, with dramatic implications for carbon dioxide emissions: "Such a decline in population growth leads to a reduction of 1.83 billion tons of carbon dioxide emissions in China per annum at present."[40]

More generally, both China and India have sought to present their "business as usual" policies as being aggressive climate policies that would remove any need to take on other obligations. For instance, both India and China have presented scenarios of their future emissions that suggest that they have already transitioned their economies to extremely high rates of decarbonization. In 2009 India released five business-as-usual projections including different assumptions of annual rates of decarbonization, from 1.0 percent up to 3.3 percent.[41] Four of these greatly exceed the 1987–2006 average annual rate of decarbonization of 1.1 percent. A single business-as-usual projection from the Chinese government released in 2009 suggested an annual rate of decarbonization of 6.5 percent per year to 2030, which is almost three times the 1987–2006 average.[42] China's emissions grew by 12.2 percent per year from 2000 to 2007, but under China's "business-as-usual" scenario growth is projected at only 2.5 percent per year to 2030.

If India and China have indeed already implemented policies that will decarbonize their economies by 3 percent per year and more, then it would be very good news indeed, as global rates of about 5 percent (or more) per year would be necessary to stabilize carbon dioxide concentrations at low levels, assuming modest economic growth.[43] However, some observers are rightfully skeptical about such claims. For example, the U.S. Energy Information Agency projects China's carbon dioxide emissions to double from its 2006 value to 12 Gt by 2030, whereas China's scenario projects an increase to only 7.5 Gt.[44]

With both India and China seeking to secure energy resources (of all types, including carbon-intensive energy resources) around the world, it seems highly unlikely that these countries have somehow discovered a secret to low-carbon growth that has escaped the United Kingdom, Japan, Australia, or the United States. With annual GDP growth expected to be 7 percent or higher per year in both countries, rapidly increasing carbon dioxide emissions seem a virtual certainty from China and India for years to come. To the extent that Brazil, Mexico, and the 193 other countries outside the top 20 emitters also seek rapid rates of economic growth, securing reliable energy supply will remain a priority with a focus on whatever energy supply can be secured at the lowest cost and greatest reliability.[45] This all but certainly will mean a continued reliance on carbon-intensive energy sources and rapidly increasing emissions from countries with large populations but relatively low emissions, such as Pakistan, Turkey, Indonesia, and Nigeria, as GDP growth continues.

The last major bloc of countries with significant emissions to discuss is Europe. Climate policy has been a core focus of policies of the European Union and many of its nations. Most notably, Europe has championed the 1997 Kyoto Protocol, which focused on reducing emissions below a 1990 baseline among industrialized countries. European officials and others (especially advocates for emissions trading) have argued that the Kyoto Protocol has been a success, resulting in lower emissions than might have otherwise occurred.[46] Others argue that the Kyoto Protocol has been a distraction. For example, Atte Korhola, professor of environmental change at the University of Helsinki, and Eija-Riitta Korhola, a member of the European Parliament, have argued that the

EU's climate policy "is expensive and flashy, yet bureaucratic and lacking results."[47]

Sorting through such claims and counterclaims about the successes or failures of Kyoto can be difficult, at best. But from the analysis in Chapters 3 and 4 we now know any policy focused on meeting aggressive emissions-reduction targets necessarily must result in an accelerated pace of decarbonization if it is going to contribute to meeting low stabilization targets. Decarbonization in the EU occurred at an annual average rate of 1.35 percent per year in the nine years before the Kyoto Protocol and 1.36 percent in the nine years following, suggesting that whatever effects the Kyoto Protocol may have had, accelerating decarbonization was not one of them during its first decade.[48] So while there are legitimate debates about what effect the protocol may have had on emissions and the degree to which counting reflects explicit acknowledgment of what was historically called "background" decarbonization, it seems unambiguous that through 2006 at least, the Kyoto Protocol did almost nothing to accelerate historical rates of decarbonization of the EU, much less raise those rates to levels needed to secure deep emissions cuts.

In many respects, climate policy is well suited to appeal to European geopolitical interests. With low rates of population growth (and population decline in some countries) and low economic growth, it is relatively much easier for Europe to achieve emissions reductions than it is for countries with high rates of population growth (like the United States) or fast-growing economies (like China or India). For Europe, business as usual results in declining emissions, especially when measured against a 1990 baseline, when emissions were much higher in grossly inefficient East Germany and before the UK "dash for gas." In 2006 David Miliband, UK secretary of state for environment, food, and rural affairs, explained why climate policy was a matter of EU interest:

> Europe needs a new *raison d'être*. For my generation, the pursuit of peace cannot provide the drive and moral purpose that are needed to inspire the next phase of the European project. The environment is the issue that can best reconnect Europe with its citizens and re-build trust in European institutions. The needs of the environment are

> coming together with the needs of the EU: one is a cause looking for a champion, the other a champion in search of a cause. . . . Climate change is the greatest challenge facing the world today. It cannot be met without the EU playing a leading role. The need to meet that challenge has the potential to bind European citizens together.[49]

But in important respects Europe has been no different from the United States, China, India, Japan, or any other country when it comes to sustaining economic growth while accelerating decarbonization—it has yet to figure it out. The iron law of climate policy holds as strongly in Europe as it does anywhere else. For instance, a spokesman for German chancellor Angela Merkel explained in 2008 why Germany wanted exemptions for certain industries from obligations to reduce emissions: "We've got to prevent companies from being threatened by climate-protection requirements."[50] In France in late 2009 a court found a proposed carbon tax unconstitutional because it exempted 93 percent of France's industrial emissions—the exemptions being necessary to win political support.[51] Following a subsequent defeat of the governing party in regional elections during the spring of 2010, French president Nicolas Sarkozy withdrew the proposed carbon tax altogether. When the trade-off is emissions reductions versus economic growth, the economy wins every time. Europe has demonstrated admirable diplomatic and symbolic leadership on climate policy, and its efforts to implement the Kyoto Protocol provide a valuable body of practical experience. Nevertheless, Europe's experiences mirror those around the world.

The bottom line from this survey of decarbonization policies around the world is straightforward: no one knows how fast a large economy can decarbonize, much less the entire global economy. Efforts to implement decarbonization policies will be better off by realizing this uncomfortable reality.

Magical Solutions and Their Consequences

The discussion and analysis in this chapter have thus far largely ignored the various and complex mechanisms of climate policy, such as those embodied in emissions trading, carbon taxes, or other instruments. The simple math of decarbonization illustrated through the preceding brief global tour shows clearly that whatever mechanism is proposed, it all but certainly cannot achieve the aggressive short-term targets set forth in climate policies in countries around the world. Rather than serving as policy targets against which politicians expect to be held accountable, emissions-reduction goals are thus to be viewed as aspirational targets that set forth a desirable but practically unachievable goal, like ending poverty or achieving world peace.

Some believe that aspirational targets are useful because they orient action in a desired direction, regardless of the pace of change. However, a risk of proposing aspirational goals is that policy makers will look for ways to avoid meeting the objectives while maintaining the appearance of accountability to formal goals, at least during their time in office. Stanford's David Victor explains the risk in the context of international climate policies: "Setting binding emission targets through treaties is wrongheaded because it 'forces' governments to do things they don't know how to do. And that puts them in a box, from which they escape using accounting tricks (e.g., offsets) rather than real effort."[52] In other words, policy makers will look to "magical solutions" that have symbolic effects but little else.

The "magical solution" to reducing carbon dioxide (and other greenhouse gas) emissions that has received the most attention is emissions trading, often known as cap and trade. Cap and trade operates under a seductively simple mechanism. Permits or allowances to emit are issued in some manner (e.g., through an auction or given away), and a market is created to allow them to be traded. A limit is set on the number of allowances available—the cap—which declines over time to some targeted value, such as a 17 percent reduction by 2020 or 80 percent by 2050. The cost of the traded allowances places a price on emissions that is set by the market, and as allowances become more scarce, the price will rise, encouraging innovation in energy technologies leading to declining carbon and energy intensities. Such trading, it is argued, will enable emissions reductions to take place where they are most efficient, as determined by the market mechanism.

Cap and trade sounds great. The problem is that it cannot work. It cannot work because it runs smack into the iron law of climate policy. As argued in Chapter 2, when emissions reductions run up against economic growth, economic growth will win out. From the perspective of the Kaya Identity—which describes the interplay of emissions, the economy, and technology—we can see that if we do not have all the technologies we need to quickly accelerate rates of decarbonization of the economy, the only other driver of emissions reductions is a reduction in GDP. Yet if a reduction in GDP is not politically possible, then what necessarily must give way is the commitment to reducing emissions. This logic means that emissions will continue to rise, even in the presence of a cap-and-trade program if technologies are not ready at scale to rapidly accelerate decarbonization.

Indeed, any effort to put a price on carbon, whether by a tax or via a cap-and-trade program, will face the same problem. Putting a high price on carbon causes economic pain and discomfort to energy consumers, who also happen to be citizens and, often, also voters. Politicians who want to continue in their jobs spend every waking hour trying to protect their constituents from economic pain. They will not rush to cause it intentionally. To think that politicians are going to willingly impose discomfort or pain on their constituents is fanciful at best.

The only way for a binding cap on emissions to not cause economic discomfort is if cap-and-trade programs are designed intentionally to have a loose or nonexistent cap, to allow economic growth to continue unaffected by the program—what might be called a nonbinding binding cap. A popular mechanism for loosening an emissions cap is through the use of "offsets," which are allowances introduced into a trading system through the reduction of emissions (or future emissions) in some distant geographical or economic location, allowing business as usual to proceed at home.

For instance, in 2009 Germany's environment minister, Sigmar Gabriel, explained that Germany needed eight to twelve new coal plants in order to meet demand while closing much hated nuclear power stations. Of the increased carbon dioxide Gabriel explained that through emissions trading, "You can build a hundred coal-fired power plants and don't have to have higher carbon-dioxide emissions."[53] Emissions trading, it seems, can work magic. In the United States congressman Rick Boucher (D-VA) expressed a similar preference for magical solutions when explaining how cap-and-trade legislation would secure a future for coal: "We provide two billion tons of offsets each year during the life of the program . . . [to be used by utilities] in forestry, agriculture and projects like tropical rain forest preservation in order to meet their carbon-dioxide reduction requirements under legislation. Therefore, they can comply with the law while continuing to burn coal."[54] Similarly, we saw Penny Wong, Australia's climate change minister, explaining earlier in this chapter how offsets would allow Australia to meet its targets for emissions reductions.

If so-called carbon offsets only allowed evasion of emissions-reduction targets, they would be bad enough. However, offsets have deeper problems. For instance, a waste product called HFC-23 results from the production of an industrial chemical used in air conditioners and some plastics.[55] HFC-23 is also a very potent greenhouse gas. Companies in China and India discovered that they could be paid by Europeans (under a Kyoto Protocol program called the Clean Development Mechanism, or CDM) to destroy the gas, which is easy to do and inexpensive. Perversely, according to Michael Wara of Stanford University, "the sale of carbon credits generated from HFC-23 capture is far more valuable than production of the refrigerant gas that leads to its creation in the first place," which had the effect that "refrigerant manufacturers were transformed overnight" into carbon-credit manufacturers with a side business in industrial chemicals. While the HFC-23 scam was identified and steps were taken to correct it (after some €4.7 billion were transferred from Europe), other perverse outcomes from emissions trading routinely surface. For instance, in late 2009 the Chinese government was accused of manipulating wind-farm subsidies so that the projects would be eligible to receive investment from Europe and generate carbon credits.[56] Many of these projects would have occurred without the European investment of more than $1 billion, despite the fact that the explicit goal of the CDM is to encourage the pursuit of less-carbon-intensive projects that would not have been built otherwise.

Even with the many failures, inefficiencies, and outright corruption demonstrated to result from cap-and-trade programs, they are unlikely to disappear anytime soon from the climate-policy landscape. Cap-and-trade advocates have invested an enormous amount of social and political capital into carbon trading. While the dismal outcome of Copenhagen in December 2009 represented a setback, for many advocates it was simply cause to try yet again. A second reason why carbon trading is not going away is more fundamental: there is an enormous amount of money involved, with an almost unlimited potential for carbon traders to make huge profits whether emissions actually go up or down. As we have seen, economic incentives are a powerful motivator. A final reason cap and trade is unlikely to go away is that some involved in the international process care more about promises than actual performance. When asked if it mattered whether Australia had passed emissions-trading legislation in time for the 2009 Copenhagen meeting, the head of the United Nations Framework Convention on Climate Change responded: "Quite honestly, no. What people care about in the international negotiations is the commitment that a government makes to take on a certain target."[57] When the focus is exclusively on ends to be achieved, the fidelity of the means employed can easily be overlooked, and magical solutions are the result.

The approach of setting an emissions target and timetable, allocating emissions permits, and then saying that the magic of the market will efficiently take care of the task is exactly the sort that one would expect if one doesn't have a good answer to the challenge of decarbonization. Markets cannot make the impossible possible, and when they are used in such a manner, they often have undesirable results.

Lessons Drawn from Decarbonization Mathematics

The bottom line of Chapters 3 and 4 is that no one really knows how to accelerate the decarbonization of large economies. The various comprehensive policies that have been put into place and proposed are clearly not up to the task, based on some very simple mathematics. One reason for this outcome is an inability to recognize those assumptions that many people "know for sure, but just ain't so" (see Chapter 2). The implications of this uncomfortable reality are not to throw up one's hands and give up. Far from it. The implication is that climate policy must proceed starting with a clear-eyed view of our policy ignorance. In such a context policy progress with respect to goals is most likely to occur with a diversity of policies that are incremental, carefully evaluated with successes scaled up and failures terminated. The design of such climate policies that might perform better is a subject that I'll return to in Chapter 9.

The climate issue is full of various authorities proclaiming this or that. Why is my argument any different? Why should you believe me? The short and simple answer is that you should not just believe what I say. You should do the math yourself. And based on the data and simple methods described in these two chapters, now you can.[58]

Figures 4.11 and 4.12 show the energy consumption mix for the top 20 global emitters in 2006 based on data provided by the U.S. Energy Information Agency and the European Environment Agency.[59] Using this information one can easily calculate total emissions for each country based on the carbon intensities of the different fuel sources. For 2006 this simple method of calculating emissions can reproduce the 2006 EIA country aggregates for the top 20 emitters to within less than a 2 percent error. These data then allow one to perform a wide range of sensitivity analyses related to how nations might hypothetically change their consumption of energy.

Table 4.4 shows the equivalent energy generation necessary to replace 10 percent of consumption for each of the top 20 countries as well as for the other 193 other countries, in terms of nuclear power stations (like Dungeness B in Kent, England), solar thermal plants (like Cloncurry in Queensland, Australia), and wind turbines (of the type being installed in West Texas).[60] For instance, the table shows that replacing 10 percent of Iran's 2006 energy consumption would require more than 11 new nuclear power stations or more than 2,500 solar thermal plants or more than 10,000 wind turbines. A reduction in consumption of 10 percent would have the same effect. One is quickly jarred back to reality when one considers the geopolitics of nuclear energy in the context of Iran. How is Iran to decarbonize?

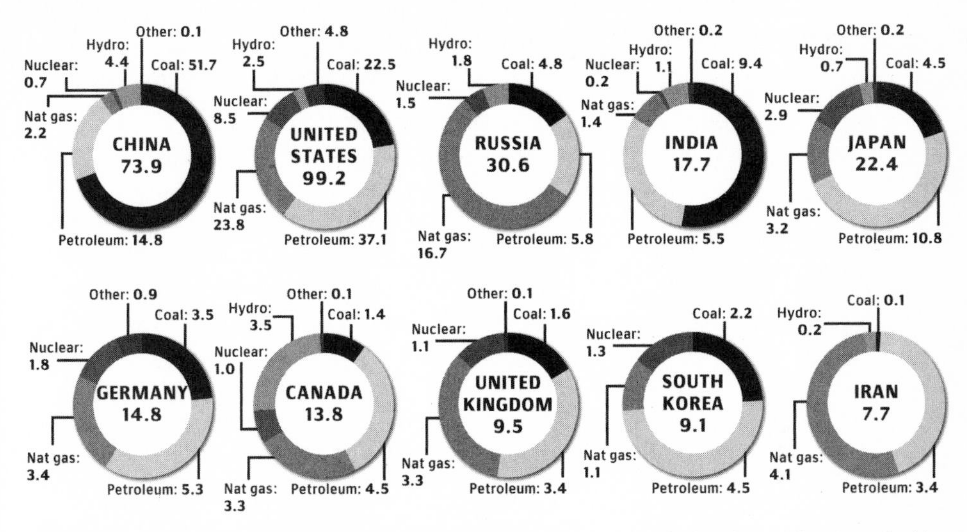

FIGURE 4.11 Breakdown of energy consumption (quads), top 1–10 global emitters of carbon dioxide.

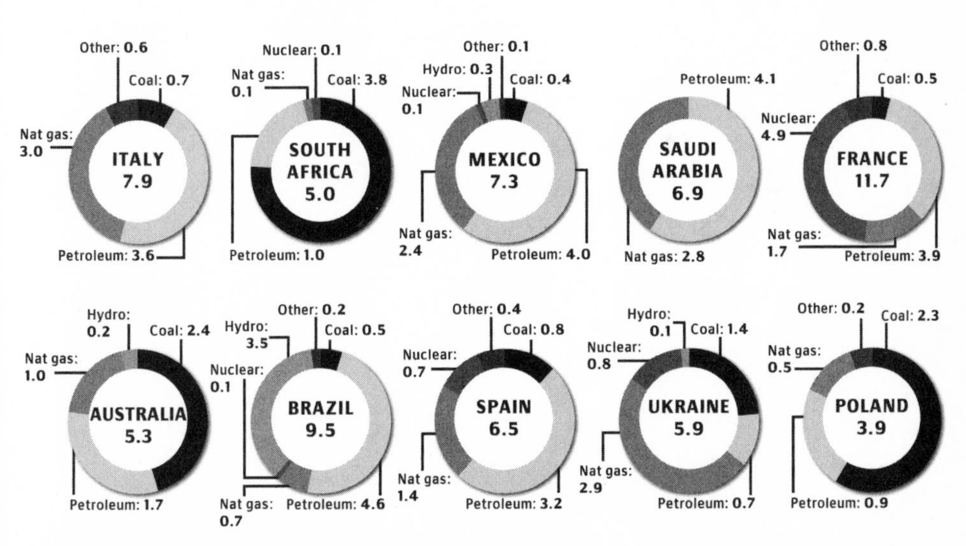

FIGURE 4.12 Breakdown of energy consumption (quads), top 11–20 global emitters of carbon dioxide. Source: U.S. Energy Information Administration and European Environment Agency.

These data can also be used to develop scenarios for emissions reductions. For instance, for the world to achieve a 50 percent reduction in its emissions below a 1990 baseline it could do the following. First, the world would need to eliminate all coal and natural gas consumption in 2006 and replace it with nuclear power stations. This could be

TABLE 4.4 Equivalent energy infrastructure

				Equivalent Energy Generation		
	Quads of Energy Consumed 2006	*10 percent of 2006 Consumption*	*10 percent of 2006 consumption in Gigawatts*	*Nuclear Plants*	*Solar Thermal Plants*	*Wind Turbines*
1 China	73.8	7.4	81.2	108.2	24,600	97,807
2 United States	99.2	9.9	109.1	145.5	33,067	131,470
3 Russia	30.3	3.0	33.3	44.4	10,100	40,157
4 India	17.7	1.8	19.5	26.0	5,900	23,458
5 Japan	22.6	2.3	24.9	33.1	7,533	29,952
6 Germany	14.6	1.5	16.1	21.5	4,876	19,388
7 Canada	14.0	1.4	15.3	20.5	4,650	18,488
8 United Kingdom	9.8	1.0	10.8	14.4	3,267	12,988
9 South Korea	9.0	0.9	9.9	13.2	3,000	11,928
10 Iran	7.7	0.8	8.5	11.3	2,567	10,205
11 Italy	8.1	0.8	8.9	11.8	2,690	10,694
12 South Africa	5.0	0.5	5.5	7.3	1,667	6,627
13 Mexico	7.4	0.7	8.1	10.8	2,452	9,750
14 Saudi Arabia	6.9	0.7	7.6	10.1	2,297	9,133
15 France	11.4	1.1	12.6	16.8	3,815	15,168
16 Australia	5.3	0.5	5.8	7.8	1,767	7,024
17 Brazil	9.6	1.0	10.6	14.1	3,212	12,769
18 Spain	6.5	0.7	7.2	9.5	2,170	8,628
19 Ukraine	5.9	0.6	6.5	8.7	1,967	7,819
20 Poland	3.9	0.4	4.3	5.7	1,300	5,169
Other 193 countries	103.3	10.3	113.6	151.5	34,438	136,922
2006 World total				692.3	157,333	625,542
2030 added demand (at 1.5 percent annual demand increase)	206.0	20.6	226.6	302.1	68,667	273,012

Information on equivalent energy generation

	Gigawatts			
Nuclear Plant	0.75	1 GW at 75 percent efficiency	Dungeness B	Kent, England
Solar Thermal	0.0033	10 MW at 30 percent efficiency	Cloncurry 10MW	Queensland, Australia
Wind Turbine	0.00083	2.5 MW at 30 percent efficiency	West Texas 2.5 MW	

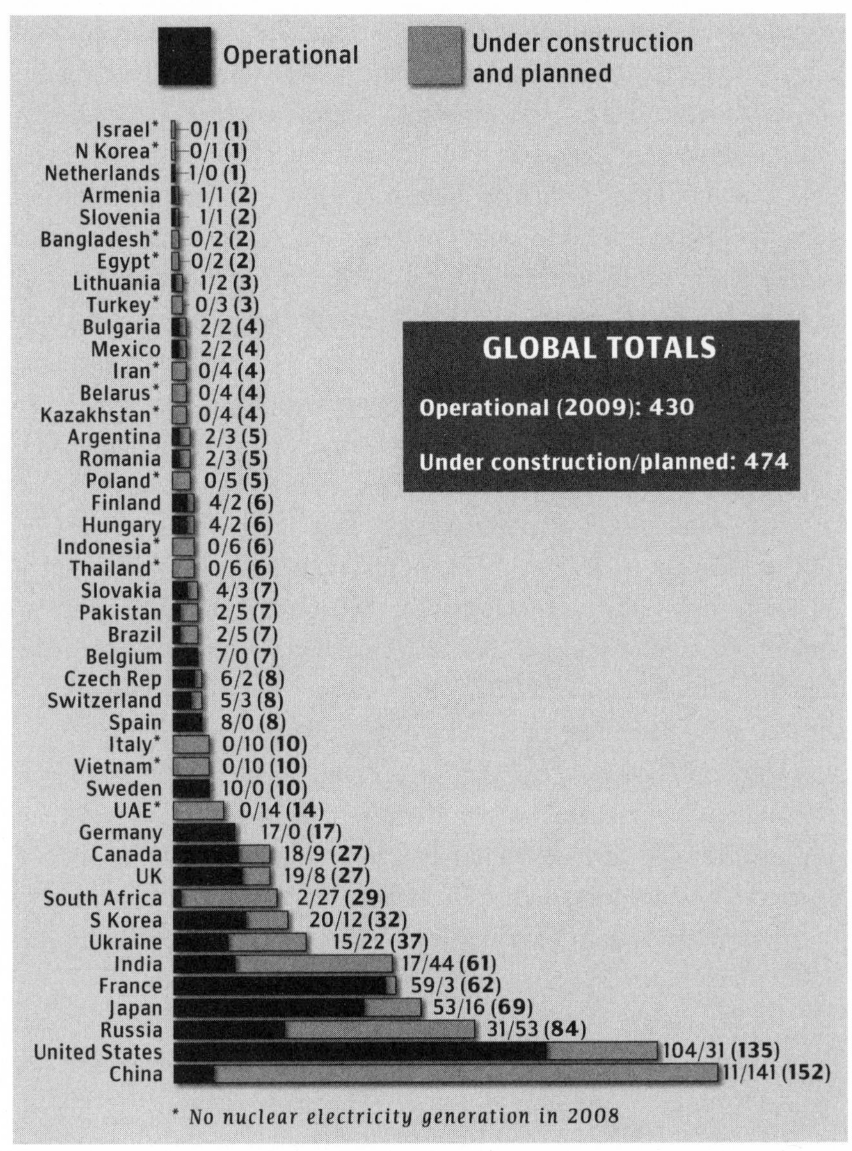

FIGURE 4.13 The global nuclear family: Distribution of reactors. Source: *Financial Times.*

done by adding about 2,800 new nuclear power plants. But that would not be enough to meet the target. More than 40 percent of 2006 petroleum consumption would also have to be replaced (e.g., perhaps by using electric vehicles), necessitating about another 750 nuclear power stations. But then there will be new demand beyond 2006 that has to be met. If global consumption of energy increases by 1.5 percent per year to 2050, that will imply a need for more than 340 new quads of energy, which, if met by nuclear power plants, implies another 5,000 nuclear power stations. The grand total? More than 12,000 nuclear power stations' worth of effort would be needed to reduce emissions to 50 percent of their 1990 level by 2050.[61] If we were to add in consumption needed to provide electricity to the 1.5 billion people in 2009 without access, it would necessitate the equivalent of thousands more nuclear stations.

These numbers are so large as to still remain a bit abstract. Figure 4.13 shows the total number of nuclear stations operating or in the planning stages as of 2009. Creating sufficient carbon-free energy by 2050 to reduce emissions by 50 percent below 1990 levels requires a level of effort equivalent to dozens of times greater than has been invested in nuclear energy to date. How many nuclear power stations is 12,000? It is, in round numbers, about the same as one new plant coming online every day between now and 2050, a result that is not new; climate scientist Ken Caldeira and his colleagues made that argument in 2003.[62]

A clear-eyed look at the simple mathematics of decarbonization and emissions reductions can be sobering, but also revealing. It need not imply that the task of accelerating decarbonization is impossible; rather, it sets the stage for a more realistic consideration of policies that might work better than those that have dominated the climate debate. But before further engaging issues of policy design for decarbonization, we have to ask: What if the decarbonization challenge proves too great? What then? Is there a backstop or Plan B? That is the subject to which we next turn.

CHAPTER 5

Technological Fixes and Backstops

THE DISCUSSION SO FAR suggests that policies now being contemplated by governments around the world to decarbonize their economies in coming years and decades are almost certainly going to fall far short of their goals. What happens if it turns out that despite the best intentions and effort, concentrations of carbon dioxide continue to increase to levels that policy makers and the public deem to be unacceptable? Recent discussions of climate policies have increasingly emphasized "geoengineering" of the global Earth system.

In January 2009 *The Independent*, a newspaper in the United Kingdom, asked eighty climate experts if the dismal performance of mitigation policies meant that a "Plan B" was now needed. The "Plan B" referred to by *The Independent* was "research, development and possible implementation of a worldwide geoengineering strategy." More than half responded in the affirmative.[1]

Geoengineering has come to mean a range of different things, and pinning down a definition is an important first step to deciding whether it's something we ought to pursue. In 2009 the American Meteorological Society defined geoengineering as "deliberately manipulating physical, chemical, or biological aspects of the Earth system" with a focus "on large-scale efforts to geoengineer the climate system to counteract the consequences of increasing greenhouse gas emissions." The AMS recognized that geoengineering overlaps with policies focused on both adaptation and mitigation: "To the extent that a geoengineering approach improves society's capacity to cope with changes in the climate system, it could reasonably be considered adaptation. Similarly, geological carbon sequestration is considered by many to be mitigation even though it requires manipulation of the Earth system."[2] Efforts to use social and economic policies to modulate concentrations of greenhouse gases might also be considered a form of "geoengineering," albeit differentiated from other attempts to manipulate the Earth system in the means employed: social policies rather than technological ones. Nevertheless, the focus in this chapter is on technological approaches to deliberately manipulating the Earth system in order to cope with or remediate accumulating carbon dioxide in the atmosphere.[3]

A wide range of different geoengineering technologies is commonly discussed in the climate debate. Perhaps the most frequently discussed approach is the injection of particles, or aerosols, into the stratosphere in order to reflect sunlight away from Earth, creating a cooling effect. The mechanism behind this proposal mimics what has been observed when a powerful volcano erupts, sending ash high into the atmosphere, such as when Mount Pinatubo erupted massively in the Philippines in 1991. The resulting effect is a temporary cooling of Earth due to the fact that the particles high up in the atmosphere reflect sunlight back into space, serving as a sort of sunshade. The effect is temporary, as the particles settle out of the atmosphere. Any effort to artificially inject particles into the stratosphere would thus have to continuously replenish the aerosols.

Stratospheric injection of aerosols was popularized by Steven Levitt and Stephen Dubner in their book *Superfreakonomics*. They wrote about a science fiction–like proposal to inject particles into the stratosphere via a giant hose tethered to a balloon, blocking sunlight a bit and cooling the earth. They proposed that such a technology could be implemented relatively inexpensively, compared to efforts to reshape the global economy in a less carbon-intensive manner. The discussion prompted a vigorous and at times acrimonious debate over geoengineering and its role in the climate debate.

Other widely discussed approaches include seeding the oceans with iron in an effort to increase the absorption of carbon dioxide into the oceans through biological processes. Such proposals have been controversial because of their unknown effects on ocean ecosystems. In 2009 a group of German and Indian scientists dumped six tons of iron into the ocean near Antarctica and evaluated its effects on taking up carbon dioxide.[4] The results were unexpectedly modest, and the experiment generated a range of opposition. Another geoengineering proposal involves the large-scale creation of clouds in marine environments, sometimes called marine-cloud brightening or whitening, in order to increase the amount of sunlight reflected back into space, and thereby cool our planet.

This chapter focuses on several different geoengineering proposals that are distinguished into two broad categories. The first category includes those technologies that seek to counteract or offset effects of increasing greenhouse gases, such as by injecting aerosols into the stratosphere. I argue that such efforts to manage the climate system are fraught with so much uncertainty and ignorance that unintended consequences are inevitable. Such efforts to "play God" in the Earth system would be confronted with technical, political, and social obstacles, making climate management a fairly undesirable approach to dealing with accumulating carbon dioxide in the atmosphere. The second category involves efforts to remove carbon dioxide from the atmosphere and to store it somewhere, which I call carbon remediation. Such technologies are unproven and expensive, and some are fanciful. However, with respect to practicality and political expediency the capture and sequestration of carbon dioxide is far preferable than those technologies that seek to deal with the consequences of increasing concentrations of carbon dioxide. While it is true that technology will be at the core of any successful effort to deal with accumulating carbon dioxide in the atmosphere, with geoengineering as with more conventional approaches to mitigation, there is (unfortunately) no quick and easy technological fix.

The Ideological Benefits and Limitations of the Technological Fix

In a 1966 lecture physicist and science policy expert Alvin Weinberg brought the concept of the "technological fix" into contemporary discussions of policy. He recognized that social problems—such as those related to population growth, education, urbanization, and environmental degradation—were “more complex” than the narrow technical challenges of building a nuclear reactor or bomb or even traveling into space. Weinberg explained that “to solve social problems one must induce social change—one must persuade many people to behave differently than they have behaved in the past.” By contrast, he notes, technological engineering was simple, and he cited the Manhattan Project to develop a nuclear bomb as an example. Given this stark difference, Weinberg asked a provocative question: “To what extent can social problems be circumvented by reducing them to technological problems?”[5]

The appeal of the technological fix is inescapable and consequently shapes our everyday lives across the range of human experience. For instance, if our eyesight is imperfect, we can have it improved with laser surgery. If we do not have enough readily available water, we can dam rivers. To ward off deadly disease, we can vaccinate children. We can, of course, adapt to poor eyesight, uncertain water availability, and extensive childhood mortality due to disease, but when such choices are available we typically choose not to, preferring to employ a technological fix. Correspondingly, wouldn’t it be great if there was a quick and easy technological fix to address accumulating carbon dioxide in the atmosphere?

Such thinking is common in the context of the contemporary climate debate. For instance, Sir Richard Branson, British entrepreneur and aviation executive, explained in the fall of 2009 that “if we could come up with a geoengineering answer to this problem, then [the 2009] Copenhagen [climate conference] wouldn’t be necessary. We could carry on flying our planes and driving our cars.” Bjørn Lomborg, the “skeptical environmentalist,” suggests that geoengineering might potentially serve as a substitute for emissions reductions: “We have a problem, and we’ve got some different options. One—climate engineering—appears to be incredibly cheap but potentially very effective, while another—carbon cuts—appears to be very expensive but not particularly effective.”[6] For some, the idea that there is a technological fix is just as appealing as the idea of a vaccine to stave off disease or laser eye surgery to correct bad vision.

Those who oppose geoengineering are every bit as committed to their views as are its supporters. For some, geoengineering is undesirable because it is in fact *societal* change that they seek to motivate via the issue of climate change. This ends-as-means type of thinking can be found in just about every complex policy issue, but it is particularly common in the climate debate. The issue of climate change has been used as a vehicle to advance causes as varied as motivating investments in alternative energy to giving purpose to the European Union, issues that had well-established constituencies long before climate change was deemed to be important. For instance, in 1988 Senator Tim Wirth, a Democrat from Colorado, argued, "What we've got to do in energy conservation is try to ride the global warming issue. Even if the theory of global warming is wrong, to have approached global warming as if it is real means energy conservation, so we will be doing the right thing anyway in terms of economic policy and environmental policy."[7]

In April 2007 a New Jersey newspaper explained how an energy expert with the advocacy group Greenpeace saw a problem with technological fixes related to climate: They "will not encourage people to develop alternate, renewable technologies, and strive for energy efficiency. Such techno-fixes also miss the point of the environmental degradation brought on by the use of fossil fuels. Carbon scrubbers won't stop oil spills, habitat-destroying strip mining and ozone."[8] Edward Parson of the University of Michigan expressed a similar view—one that is widely shared among many advocates in the climate debate—that social change is a desired end in addition to addressing accumulating carbon dioxide in the atmosphere: "Can we continue to rely on narrowly drawn policies to motivate reductions in whatever specific forms of production and consumption pose the clearest immediate risks? Or must we somehow find a way to address the larger-scale question of limiting the aggregate scale of human population and economic activity, and seek to identify some means to achieve this that is compatible with humane, democratic states that value individual liberty?" From this perspective the technological fix may address some symptom of a problem but leaves intact those conditions that are viewed to be the problem itself. Of course, people strongly disagree about the nature of the problem.[9]

The debate between geoengineering supporters and its opponents helps to reveal what should be obvious: addressing the accumulating concentrations of carbon dioxide is not the same thing as addressing climate change. For some, carbon dioxide is a mere symptom of a world with its values out of kilter. Addressing climate change thus means much more than can ever be addressed via a technological fix, whether that fix involves geoengineering or technologies more mundane, such as alternative energy supply. For others, a bit of excess carbon dioxide in the atmosphere and the oceans is just another small annoyance of the sort that inevitably results from the inevitable progress of the human race. Like other annoyances, such as the problems of acid rain or ozone depletion, they might argue, we'll figure this one out as well, and geoengineering offers an obvious solution. Writing forty-five years ago, Weinberg was quite aware of these two "ways of dealing with a complex social issue: The social-engineering way asks people to behave more 'reasonably,' the technologist's way . . . tries to avoid changing people's habits or motivation."[10]

These disparate perspectives are inherently irreconcilable, as they are rooted in vastly different ideological perspectives about progress, humanity, nature, and even religion. Of course, those holding these different perspectives do not often agree on solutions, since they are far from even agreeing on the nature of problem. Mike Hulme has written eloquently about climate change as a "magnifying glass and as a mirror." As a magnifying glass it enables us to examine the consequences of our short-term decisions for how we wish to live. As a mirror it forces the values that we share and do not share into the open. Hulme explains, quite convincingly in my view, that "climate change" is a sort of "wicked problem" that lacks anything resembling a solution, and the very best that we can do is to employ "clumsy solutions" that are a sort of muddling through. We can do better or worse in response to the problem, but we cannot solve it, not least because different people are far from agreeing as to what that problem actually is. As Weinberg writes, "Technology will never replace social engineering," but it can certainly be useful, as anyone can attest who enjoys corrected eyesight, water in a dry region, or healthy children.

So instead of "climate change" writ large, what if instead we considered geoengineering from the perspective of the somewhat tamer problem of accumulating carbon dioxide in the atmosphere? What might the role of technological fixes be in that context? This, of course, is not to deny the broader issues at stake in the climate debate, nor is it to gloss over the fact that accumulating carbon dioxide is not equivalent to climate change, a topic that I'll discuss in some depth in Chapter 6. However, even while granting that addressing carbon dioxide would not address important issues of equity, development, and economy, addressing the accumulation of carbon dioxide in the atmosphere has been accepted on its own terms by many advocates and special interests as something worth doing, as explained in Chapter 1. It is therefore worth investigating the role of technological fixes to that end, stepping away from the ideological debate about climate change considered more broadly.

Criteria for a Successful Technological Fix

Does geoengineering offer hopes of a quick and effective technological fix in response to accumulating carbon dioxide in the atmosphere? To evaluate the potential of any technological fix requires us to apply some criteria to help discern those proposals likely to succeed and those likely to fail. Writing in *Nature* in December 2008, Dan Sarewitz and Dick Nelson offered three criteria by which to distinguish "problems amenable to technological fixes from those that are not."[11] Sarewitz and Nelson's analysis provides an extremely useful and straightforward guide to understanding those contexts in which a proposed technological fix holds promise and those in which it does not.

Criterion 1: The Technology Must Embody a Cause-Effect Relationship

The technology should do what it is advertised to do. A vaccine should prevent disease, laser eye surgery should improve eyesight, a dam should reliably store water. It seems fairly obvious that technologies that don't do the work they are supposed to cannot be claimed to be useful "fixes." Any technology that is somehow supposed to manage the effects of or remediate accumulating carbon dioxide emissions should be judged on that basis.

Criterion 2: The Effects of the Technological Fix Must Be Assessable

When deployed the effects of the technology should be identifiable. Meeting this criterion was a significant problem for policies focused on weather modification, deployed at various times and locations beginning in the 1950s in efforts to manage rainfall and steer hurricanes. When scientists conducted weather-modification experiments, such as cloud seeding, they had a difficult time assessing whether the resulting rainfall was caused by the modification or would have occurred anyway. Knowing the effectiveness of such interventions proved difficult from a basic scientific perspective. Geoengineering policies should be evaluated on their effects, which should be able to be conclusively attributed to the geoengineering intervention.

Criterion 3: Research and Development Must Focus on Improvement

Science is most practical when it works from an established base of knowledge for which there is much experience. Sarewitz and Nelson explain that vaccine development is based on more than two centuries of experience with its application and study. This core of knowledge contributes to the effectiveness of research and development resources invested in vaccine technology. Whatever research is done into geoengineering, we should expect that any practical applications that follow from such research will first be in areas where there is an existing body of knowledge and experience. This does not preclude significant or even revolutionary technological breakthroughs that fundamentally transform issues, but we should expect such breakthroughs to be very rare.

The following sections apply these three criteria in the context of technologies of climate management and technologies of carbon remediation, suggesting that a more general argument can be extended to other techniques of geoengineering that share similar characteristics. The first application of the three criteria focuses on technologies of climate management, which are essentially trying to play God with the climate system.

The second involves technologies of carbon remediation, which I characterize as "cleaning up our mess."[12] I conclude that technologies for climate management fail comprehensively with respect to the three criteria for a technological fix, whereas the technologies of carbon remediation offer far more promise, though with significant obstacles nonetheless.

Can (or Should) We Play God with the Climate System?

In 2006 Nobel laureate Paul Crutzen published a paper in the journal *Climatic Change* that brought discussions of geoengineering more squarely into the focus of scientific debates, and much more into public view, than they'd ever been.[13] In a subsequent comment—one that reflects the all too common practice of scientists' seeking to influence political outcomes through their science—Crutzen explained that rather than legitimating geoengineering, his paper was in fact designed to motivate efforts to reduce emissions. "It was meant to startle the policymakers," Crutzen said. "If they don't take action much more strongly than they have in the past, then in the end we have to do experiments like this."[14] The same sort of "scientific threat" was explicit in the press release that accompanied the 2009 UK Royal Society report on geoengineering that bluntly warned, "Stop emitting carbon dioxide or geoengineering could be our only hope."[15]

Not at all surprisingly, the "scientific threat" has been viewed by many not as a risk but as a potential resolution to the entire climate issue. These dynamics were fully apparent in a 2009 exercise led by Bjørn Lomborg that I participated in called the "Copenhagen Consensus on Climate Change." In principle, the point of the exercise was a worthy one. It was to bring together a group of Nobel laureates in economics to examine a range of policies proposed in response to climate change and evaluate how they compared to one another in terms of their costs and benefits. My role in the exercise was to write a paper on geoengineering responding to an evaluation of the costs and benefits of several proposed geoengineering technologies prepared by J. Eric Bickel of the University of Texas and Lee Lane of the American Enterprise Institute. Bickel and Lane focused on two technologies that would increase the amount of sunlight reflected back into space, thus creating a cooling effect: the injection of aerosols into the stratosphere and marine-cloud brightening. They argued that based on their cost-benefit analysis, geoengineering appeared extremely appealing, and they called for an investment of $750 million per year to develop and evaluate the proposed technologies. While their analysis stopped short of calling for full deployment of these technologies, subsequent discussions left little question whether the participants in the Copenhagen Consensus exercise liked the idea of geoengineering. Lomborg subsequently championed geoengineering in a series of op-eds.

It seems obvious that conducting a meaningful cost-benefit analysis requires some degree of accuracy in estimates of both costs and benefits of alternative courses of action. In the cases of stratospheric aerosol injection and marine-cloud whitening there are considerable uncertainties in direct costs of deployment, not least because real-world geoengineering experiments have never actually been done. This means that there are areas both of uncertainty and of fundamental ignorance, where even uncertainties are not well understood. Ronald Prinn, a professor of atmospheric sciences at MIT, explains the basic problem: "If we lower levels of sunlight, we are unsure of the exact response of the climate system to doing that, for the same reason that we don't know exactly how the climate will respond to a particular level of greenhouse gases. That's the big issue. How can you engineer a system you don't fully understand?"[16]

But let us assume that direct costs of the technologies (i.e., implementation at scale) are known with some degree of accuracy, such that they pose no obstacle to conducting a meaningful cost-benefit analysis. There would remain areas of fundamental ignorance in the estimates of indirect costs and potential benefits that are fatal to efforts to create a meaningful cost-benefit analysis of potential geoengineering technologies. When a quantitative analysis of any type seeks to deal with uncertainty or ignorance, then some simplifying assumptions must be adopted to enable any calculations to be made at all. Such assumptions can be made in any number of potentially plausible ways, in some cases even leading to diametrically opposed conclusions. And when the outcome of an analysis rests entirely on the choice of assumptions, without any empirical means to discriminate between them, then the exercise can do more to obscure than reveal.

When I was writing my response paper to Bickel and Lane I discovered a very similar analysis of the costs and benefits of geoengineering by scholars at Penn State University. Like Bickel and Lane, the Penn State scholars used a version of the same computer model to assess the economic and climatic effects of stratospheric aerosol injection. However, the Penn State team started with a set of different, but no less plausible, assumptions at the outset of their analysis. Contrary to the Copenhagen Consensus analysis, the Penn State analysis concluded that aerosol injection into the stratosphere "fails an economic cost-benefit test in our model for arguably reasonable assumptions."[17] I highlight the divergent results not to argue that one is more correct than another but to point out that in this case, two different research teams using very similar methods and just by varying assumptions about "deep uncertainties" arrived at results that are completely at odds with each other.

The conclusion, then, is that while it is certainly plausible that techniques of geoengineering could lead to very large benefits in relation to costs, it is also possible that those same techniques could lead to very large costs with respect to benefits. There is simply no way at this point to adjudicate empirically between these starkly different conclusions, or even to calculate meaningful probabilities of different outcomes. Lord Rees, president of the Royal Society, commented on geoengineering in a debate before the UK Parliament in early 2010, "Such techno fixes have an undoubted allure for some people, but our [recent Royal Society] report emphasized that geo-engineering could have unintended consequences, as well as being plainly politically problematic."[18] It is this fundamental, and arguably irreducible, ignorance that leads to the conclusion that "more research is needed." In 1992 a United States National Academies of Science committee assessing geoengineering strategies concluded, "Engineering countermeasures need to be evaluated but should not be implemented without broad understanding of the direct effects and the potential side effects, the ethical issues, and the risks."[19] Today, almost twenty years later, we still lack this broad understanding.

Because geoengineering research has considerable value to advancing fundamental understandings of the global Earth system, there are other justifications for its support beyond the potential development of climate-engineering technologies. As efforts to understand the impacts of geo-engineering have made clear, our conceptual model of the global Earth system is flawed and overly simplistic. Simple frameworks can be very useful tools for advancing scientific understandings, and sometimes even for policy making. The bathtub analogy for carbon in the atmosphere is an example of a simple but useful framework. However, there are other times when a simple framework can lead one dangerously astray.

Here a brief excursion from geoengineering is worthwhile. In 1935 102 cane toads were brought to Australia from Hawaii.[20] The purpose of the importation was an effort to control the cane beetle, so named because of its affinity for sugarcane. The thinking was that the cane toad would eat the beetles, helping to control their population and at the same time the fortunes of the sugarcane industry. The framework here was simple: frogs eat beetles; fewer beetles mean less damage to the sugarcane crop. Thousands of toads were ultimately let loose in northeastern Australia. The problem, of course, is that ecosystems are complex, open systems that are not fully understood—far from it. In the case of the cane toads it turned out that they favored other sources of food than beetles (a clear violation of Sarewitz and Nelson's first criterion), and with no natural predators, the population exploded. Today there are more than 200 million cane toads in Australia, presenting a serious ecological problem. Policy making is replete with lessons of unintended consequences resulting from the failure to appreciate the complexity of complex, open systems in favor of simple but potentially misleading simplifications.[21] As one recent study concluded after looking at efforts to manage complex ecological systems, "We must approach the question of geoengineering with both caution and an awareness of the lessons from past ecological interventions."[22]

Like ecological systems, the climate system is enormously complex, requiring simplifications for effective analysis, which can, however, also be misleading when it comes to efforts to intervene. Geoengineering proposals are often evaluated through a framework of "radiative forcing," which is a measure of a change in the energy budget of the Earth system due to some perturbation, such as a change in the amount of greenhouse gases in the atmosphere. The concept of radiative forcing is very appealing because it enables different perturbations of the climate system to be compared with one another on a common scale. The standard basis for comparisons of different human (and nonhuman) influences on the climate system is typically with respect to the influence of carbon dioxide.[23] Indeed, it is this metric that enables different greenhouse gases to be compared with each other in terms of their influence on radiative forcing, thereby enabling policies to address "baskets" of greenhouse gases. For instance, according to the IPCC, HFC-23 has a carbon dioxide equivalency of 14,800, which means that its radiative forcing effect is 14,800 times that of carbon dioxide.[24] Unfortunately, the concept represents an understanding of our climate system that is, at best, imperfect.

Carbon dioxide equivalencies are problematic for several reasons. Different chemicals have different residence times in the atmosphere; a long-lived greenhouse gas is potentially a bigger threat and more difficult to deal with than one that quickly breaks down in the environment. The system of carbon dioxide equivalencies also fails to account for the multiple pathways and feedbacks through which different human activities have an influence on the climate system.[25] The 2007 IPCC report noted that the concept of radiative forcing is very useful but that "it provides a limited measure of climate change as it does not attempt to represent the overall climate response."[26] The IPCC also cautions against simply summing up various radiative forcing terms to arrive at a net human influence. In 2005 the U.S. National Research Council (NRC) offered an even more explicit warning: "The assumed linearity of radiative forcing has been simultaneously useful and misleading for the policy community. It is important to determine the degree to which global . . . forcings are additive and whether one can expect, for example, canceling effects on climate change from changes in greenhouse gases on the one hand and changes in reflective aerosols on the other."[27] In other words, two different climate perturbations with opposite signs in terms of overall radiative forcing do not simply cancel each other out.

A more complex view of radiative forcing and its relationship to nonradiative forcings, indirect radiative forcings, and their feedbacks is shown in Figure 5.1. The relationship of a forcing agent, such as the injection of stratospheric aerosols or marine-cloud whitening, and eventual climatic impacts at global as well as regional scales manifests itself in a degree of interrelationships and feedbacks that cannot be characterized simply by adding or subtracting direct radiative forcings. This means that it is overly simplistic to think of geoengineering as providing a counterbalance to the effects of accumulating carbon dioxide in the atmosphere. Alan Robock of Rutgers University warns, "Scientists cannot possibly account for all of the complex climate interactions or predict all of the impacts of geoengineering. Climate models are improving, but scientists are discovering that climate is changing more rapidly than they predicted, for example, the surprising and unprecedented extent to which Arctic sea ice melted during the summer of 2007. Scientists may never have enough confidence that their theories will predict how well geoengineering systems can work. With so much at stake, there is reason to worry about what we don't know."[28] Modeling studies show that geoengineering techniques could have a wide range of undesirable effects, such as on regional weather patterns and precipitation.[29] Unintended consequences are certain.

How, then, does geoengineering in the form of seeking to counterbalance the effects of accumulating carbon dioxide with various interventions—climate management—stack up with respect to the three criteria for a successful technological fix proposed by Dan Sarewitz and Richard Nelson? Not well.

Criterion 1: The Technology Must Embody a Cause-Effect Relationship

The ability to conduct a cost-benefit analysis of climate engineering is hindered by both uncertainties and fundamental ignorance of both costs and benefits. It is possible to vary assumptions in plausible ways and to arrive at diametrically opposed results. Further, simplifications of physical relationships may be suitable for inclusion in a simple, exploratory integrated assessment model. However, if the simplifications fail to reflect the actual complexity of the global Earth system, it could

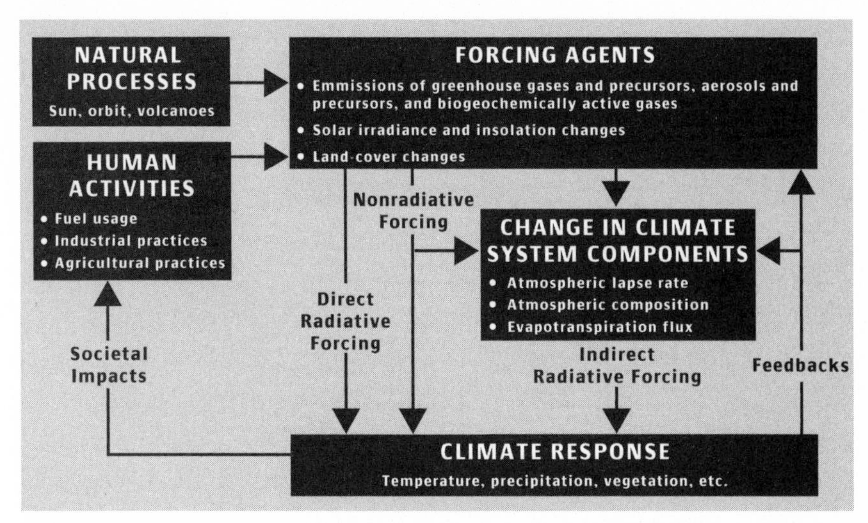

FIGURE 5.1 Radiative forcing in context. Source: National Research Council.

lead analysts astray. For instance, geoengineering techniques do not simply and directly offset the effects of accumulating carbon dioxide emissions. They have effects on the climate system through direct and indirect radiative forcing, nonradiative forcings, and various feedbacks among these. Fully characterizing these effects, much less making accurate and reliable predictions, is far beyond the capabilities of contemporary climate science and will likely remain so.

Criterion 2: The Effects of the Technological Fix Must Be Assessable

Imagine if the world decided to embark on a grand geoengineering effort by injecting particles into the stratosphere. Then assume that the Northern Hemisphere subsequently experienced an extreme winter similar to that of 2009–2010, with frigid temperatures and copious snowfalls. What would then happen?

Obviously, some would blame the geoengineering for the weather extremes. Others would argue that such winters happen from time to time. Scientifically, there would be multiple and conflicting explanations for what was observed. Efforts to modify the weather in the mid- to late twentieth century saw issues of responsibility for observed outcomes founder on causality.[30] William Travis, a disaster expert at the University of Colorado, explains that the more that atmospheric scientists learned about the complexities associated with the effects of weather modification, the more difficult it became to demonstrate causality between the modification and the resulting weather.[31] At the same time, those doing the weather modification were viewed by the public and policy makers to be responsible for whatever bad outcomes resulted from the weather, regardless of causality. It is easy to imagine geoengineering schemes resulting in similar outcomes.[32] The bottom line is that understandings of the effects of geoengineering on global weather and climate would be highly contested, with little means for resolving such controversies unambiguously and empirically through science.

Criterion 3: Research and Development Must Focus on Improvement

Geoengineering technologies do not exist. There is no practice planet Earth on which such technologies can be implemented, evaluated, and improved. Even the most sophisticated climate models are but a simplistic facsimile of the real planet Earth. There is thus no technical core on which geoengineering technologies can be developed and improved. Developing such a technical core via trial and experience is limited by the fact that we have only one planet, and criteria 1 and 2 would seem to prevent developing a technical core of sufficient quality to allow for reliable judgments of cause and effect and therefore consequences. This might seem paradoxical: one reason for opposing implementation of geoengineering technologies is that they are poorly understood, but developing such understandings is limited because of obstacles to implementation. This paradox does not seem breakable.

To summarize, geoengineering in the form of climate management fails comprehensively with respect to the three criteria for a technological fix. Whatever the effects of climate management on the climate debate or efforts to motivate societal change through the issue of climate change, climate management fails a prior step. It is not a technological fix at all—far from it. Climate management deserves to stay on the pages of science fiction novels and to remain the subject of speculative research. However, it offers no prospect for addressing the consequences of accumulating carbon dioxide in the atmosphere. We'll have to look elsewhere.

Carbon Remediation: What About Just Cleaning Up Our Mess?

It might seem perfectly obvious to observe that if accumulating carbon dioxide in the atmosphere presents a problem requiring action, then we should just take it out of the atmosphere and store it somewhere safe and permanent. This is akin to what we do, for instance, with nuclear waste and arsenic in water supplies. There are many proposed approaches to the direct removal and storage of carbon dioxide from the atmosphere, which I call carbon remediation. Removing carbon dioxide from the atmosphere offers a direct approach to dealing with accumulating concentrations in the atmosphere. In terms of the bathtub analogy, carbon remediation would involve removing water from the tub and storing it somewhere. For instance, you might use a bucket to remove water from the tub and then pour it into a nearby sink. If you remove water at the same rate as the tub is being filled, then the tub will not overflow. (Of course, the implication of the analogy is that so long as water is continuing to fill the bathtub, you will be committed to taking bucket loads of water to the nearby sink.)

In recent years the possibility of air capture of carbon dioxide has found increasing attention in prominent reports of the United Kingdom's Royal Society and Institution of Mechanical Engineers, as well as on the pages of *Science* and *Nature*. In 2006 Al Gore and Richard Branson called more attention to air capture when they announced the Virgin Earth Challenge, promising $25 million to the first "commercially viable design which results in the removal of anthropogenic, atmospheric greenhouse gases." Well-known scientists James Hansen and Wallace Broecker have also advocated air-capture technologies. However, the amount of effort being devoted to research on air capture has been quite small, with estimates of a few dozen people and a few million dollars invested into the technology worldwide. This seems likely to change.

A diverse range of carbon-remediation technologies has been proposed, and some have been implemented; these approaches include biological, chemical, and geological capture techniques. One approach focuses on the capture of carbon dioxide from either the air around us or the carbon dioxide–rich exhaust from a gas- or coal-fired power plant. The latter has received considerable attention from researchers, and policy makers around the world have already placed big bets on the future availability. But the scale of carbon-capture implementation is daunting. The *Wall Street Journal* aptly summarized the implications of a commitment to carbon capture and sequestration found in a 2009 IEA report:

> The report estimates the "additional cost" of all the carbon capture projects in the world at $5.8 trillion. . . . Then there is the sheer physical difficulty of installing 3,400 carbon-capture projects around the world by 2050: That's an average of 85 projects per year, every year, till the middle of the century, or one every four days. Starting pitchers can't even keep up that pace to throw a few innings; imagine trying to make, move, install, test, and commission large-scale carbon-capture projects at that pace. And then find a place to put all those millions of tons of carbon dioxide. . . . The carbon emissions that are caught have to get stuffed underground via pipelines. The IEA figures 360,000 kilometers of pipeline should do the trick. That's nine trips around the earth. Somebody better lock up steel futures, if that's the case.[33]

Forests are perhaps the classic example of biological capture and sequestration. Arguments to expand and protect existing forests as reservoirs of carbon are meant to avoid its release to the atmosphere and absorption by the oceans. Not surprisingly, many advocates of forest protection have found this line of argument very convincing and have consequently engaged readily in the climate issue. So, too, have many agricultural interests, who see the opportunity to benefit financially from policies that pay farmers to manage their lands in ways that enhance the uptake and sequestration of carbon.

Some have taken such proposals for biological capture and sequestration much further. NASA's Jim Hansen and colleagues proposed that carbon dioxide emitted from power plants fueled with biomass might be captured at the source and then sequestered in the deep sea or below the seafloor. Another biological approach is to create charcoal from plants (called biochar) and bury it in soil. James Lovelock, the father of the Gaia theory, has advocated biochar as a means of drawing down and storing carbon.[34] An even more speculative proposal involves using desalinization plants to turn great deserts of the world, including swaths of the Sahara and Australian Outback, into jungles, creating vast new stores of carbon.[35]

With respect to the capture of carbon dioxide from the ambient air using chemical technologies, Columbia University's Frank Zeman argues that "nobody doubts it's technically feasible," but he also points out the costs and technical challenges. Fellow Columbia University professor and air-capture entrepreneur Peter Eisenberger says that some technology is going to be necessary to draw down carbon dioxide levels: "Without having something that is carbon negative, the possibility of avoiding high levels of CO_2 is basically zero."[36] Yet, in contrast to forests and carbon capture from power plants, the direct capture of carbon dioxide from the ambient air—air capture—has received remarkably little attention in debates on policy responses to climate change, but this seems to be changing.[37]

The technology of chemical air capture is not new. It was studied as early as the 1940s and proposed in the 1970s as a source of energy, with the captured carbon recycled into gas or liquid fuels. In recent years in the context of climate change, various proposals have been advanced for air capture, several have been operationally tested, and several are being commercialized. David Keith at the University of Calgary developed a prototype system that uses sodium hydroxide and lime to remove carbon dioxide from the air. Klaus Lackner, another Columbia University professor with expertise in air capture, has proposed an alternative absorption technology that does not use sodium hydroxide, but it is not described in detail due to its proprietary nature. In early 2010 *Science* published a paper detailing a serendipitous discovery of the use of a copper-based sorbent to capture carbon from the air using less energy than previous methods.[38]

Presently, there are no experimental data on the complete process of air capture, especially at scale, to demonstrate the concept, its energy use, and the engineering costs. However, that need not preclude doing a few sensitivity analyses using simple math and assuming that these costs would hold at scale.[39] In the exercises below I use three values for the costs of air capture from the work of David Keith, Klaus Lackner, and the IPCC: $140, $100, and $27 per metric ton of carbon dioxide, using technologies that were available in 2007.[40]

Assume it costs $140 to capture one metric ton of carbon dioxide from the atmosphere. At 7.8 Gt of carbon dioxide equivalent to 1 part per million of carbon dioxide in the atmosphere, it would cost about $1 trillion to reduce the concentration of carbon dioxide in the atmosphere by 1 part per million. That sum represents about 1.6 percent of global GDP in 2009. At $140 per metric ton of carbon dioxide, the complete remediation of net 2009 human emissions (the part not absorbed by the land surface and the oceans) would cost about $3 trillion to $4 trillion, or about 5 to 7 percent of global GDP. Remediating total human carbon dioxide emissions would be about twice these values. At $100 per ton the cost to remediate total carbon dioxide emissions would be about 2.5 to 3 percent of global GDP. These are very rough estimates, of course, but they do provide a sense of scale. Interestingly, they are quite similar to the costs of conventional mitigation of carbon dioxide emissions presented in leading climate assessments.

Another way to look at the costs of air capture is with respect to the cumulative costs of removing carbon dioxide from the atmosphere over a period of time. For instance, we can use the concept of stabilization wedges to arrive at a back-of-the-envelope cost estimate. Socolow and Pacala argued that we would need to reduce emissions by seven stabilization wedges by midcentury to be on a path to stabilizing atmospheric concentrations of carbon dioxide at 500 ppm; in Chapter 2, I argued that this number could actually be as many as twenty-five or more wedges. The former implies a cumulative reduction of about 640 Gt of carbon dioxide and the latter about 2,750 Gt, a range that includes the central estimates of the IPCC presented in 2007. At air capture costs of $27 and $140 per metric ton, this provides a range of 0.5 to 9 percent of cumulative 2010–2050 global GDP, with a middle value of about 4 percent.[41]

Remarkably, this number is within the range of uncertainty for the costs of conventional mitigation as suggested by the IPCC and the 2006

Stern Review Report on climate change produced by the United Kingdom. The Stern Review explained how one might think about 1 percent of GDP: "If mitigation costs 1 percent of world GDP by 2100, relative to the hypothetical 'no climate change' baseline, this is equivalent to the growth rate of annual GDP over the period dropping from 2.5 percent to 2.49 percent. GDP in 2100 would still be approximately 940 percent higher than today, as opposed to 950 percent higher if there were no climate-change to tackle."[42] If air-capture technology could be implemented at $27 per metric ton (or less) and storage of captured carbon dioxide proved feasible, then it would likely reshape the nature of the climate debate in the direction of the possibility of a technological fix contributing at least part of the means to address accumulating carbon dioxide in the atmosphere.

Making global cost estimates for any complex set of interrelated systems far into the future is a dubious enterprise. However, the analysis here shows that by using assumptions consistent with those of the IPCC and Stern Review, air capture compares favorably with the cost estimates for mitigation provided in those reports. The main reason for this surprising result, given that air capture has a relatively high cost, is the long period for which no costs are incurred until the stabilization target is reached, while GDP continues to grow. The central value for the costs of air capture under the assumptions examined here is also less than the projected costs of unmitigated climate change over the twenty-first century, which the Stern Review estimated to be from 5 to 20 percent of GDP annually and the IPCC estimated in 2007 to be 5 percent of global GDP by 2050. The accompanying text box illustrates the costs to gasoline of using air capture to offset U.S. automobile emissions, using the cost figures presented above.

There are also several factors that serve to overstate the cost estimates of air capture presented here as well. Carbon dioxide emissions from power plants, representing perhaps as much as half of the total emissions over the twenty-first century, could perhaps be captured at the source for a cost considerably less than direct air capture. The technical, environmental, and societal aspects of carbon sequestration are identical for capture of carbon dioxide from both power plants and ambient air. To

Full Mitigation of U.S. Auto Emissions with Air Capture

One can also engage in a much more focused analysis of the costs of air capture in climate mitigation. For instance, in 2005 U.S. auto emissions were responsible for about 6 percent of total global emissions. Six percent of 2007 carbon emissions are about 0.48 Gt. All U.S. automobile emissions of carbon dioxide could be offset through air capture for a cost of $48 billion at $100 per metric ton of carbon, $173 billion at $360 per metric ton of carbon, or $240 billion at $500 per metric ton of carbon. For comparison, were the United States to have signed on to the Kyoto Protocol requiring a 7 percent reduction in 1990 levels of emissions, the annual cost of meeting this target via air capture (using 2006 emissions values) would have been about $125 billion at $360 per metric ton of carbon or about $173 billion at $500 per metric ton of carbon. In November 2007 in Europe the cost of a 2008 certified emissions-reduction credit under the Kyoto Protocol was about $100 per metric ton of carbon, so air capture is 3.6 to 5 times more expensive than the costs to signatories of Kyoto.

The values for offsetting U.S. carbon dioxide emissions from gasoline in automobiles may be easier to understand in terms of the cost of a gallon of gasoline. In 2005 the United States used approximately 140 billion gallons of gasoline. Assuming 150 billion barrels for 2007 equates to a gas tax of $1.15 (at $360 per ton) or $1.60 per gallon (at $500 per ton). These levels of taxation are smaller than gas taxes in many European countries. To place these values into domestic context, the average cost of a gallon of gasoline in the United States increased by $1.60 between December 2003 and August 2006. The average cost of a gallon of gasoline increased by $0.32 between April and May 2007. In principle, under the assumptions here, all of the emissions of carbon dioxide from automobiles in the United States could be removed via air capture using 2007 technology at marginal costs that are of the same magnitude as the interannual variability of gasoline prices, and U.S. consumers would still have among the lowest gasoline prices in the world. This example does not consider storage of the removed carbon dioxide, which would add to the costs.

Sources: "World Carbon Dioxide Emissions from the Consumption of Petroleum, 1980–2005"; Energy Information Administration, http://www.eia.doe.gov/pub/international/iealf/tableh2co2.xls; Energy-Related Carbon Dioxide Emissions, Energy Information Administration, chap. 7, http://www.eia.doe.gov/oiaf/ieo/emissions.html; Emissions of Greenhouse Gases in the United States, 2006; Office of Integrated Analysis and Forecasting, U.S. Department of Energy, DOE/EIA-0573, ftp://ftp.eia.doe.gov/pub/oiaf/1605/cdrom/pdf/ggrpt/057306.pdf; "Energy Basics 101: Basic Petroleum Statistics"; Energy Information Administration, http://www.eia.doe.gov/neic/quickfacts/quickoil.html.

the extent that improvements in efficiency and overall emissions intensity occur, these developments would further reduce total emissions and thus the need to rely on air capture. The assumptions here assume simplistically a fixed average cost of air capture over time, whereas experience with technological innovation suggests declining marginal costs over time. Consideration of these factors could reduce the values presented here by a significant (but unknown) amount. That said, uncertainties in rates of increasing emissions, economic growth, and future atmospheric concentrations of carbon dioxide mean that the values presented here could be more or less under different assumptions.

Nevertheless, the result of this analysis suggests that, at a minimum, air capture should receive the same detailed analysis as other approaches to mitigation, and even be looked at as a potential contributor to stabilizing greenhouse gas concentrations as a low-level or backstop technology. In several decades we may want to have well-developed and well-tested technologies of air capture available; we may well want a brute-force approach on hand. But the technology for such an approach is unlikely to appear unless there is a focused effort on innovation in the direct capture and storage of carbon dioxide, an effort that to date does not really exist on any appreciable scale.

So how do technologies of carbon remediation stack up with respect to the three criteria for the success of technological fixes? Some technologies of carbon remediation have potential, others much less so.

Criterion 1: The Technology Must Embody a Cause-Effect Relationship

Technologies that directly remove from the atmosphere and permanently store carbon dioxide are extremely attractive because they offer a direct approach to addressing accumulating carbon dioxide in the atmosphere. Thus, more so than any other approach, capture and sequestration embodies the cause-effect relationship that lies at the core of concerns relating to carbon dioxide.

At the same time, because of the directness in the cause-effect relationship, capture and sequestration offers very little benefit to those who see accumulating carbon dioxide as a symptom of a deeper problem or even as a vehicle to address related concerns, such as overconsumption or population growth. Even so, a focus on capture and storage is conceptually useful in the debate because it can help to distinguish various means and ends of various advocates for different sorts of action.

Criterion 2: The Effects of the Technological Fix Must Be Assessable

Another compelling reason that capture and storage makes sense is that it is straightforward to assess the amount of carbon dioxide captured and the fidelity of any storage reservoir. Unlike other forms of geoengineering, there would be little need to speculate about secondary efforts or unexpected consequences. The success or failure of capture technologies could also be evaluated at small scales before being widely deployed.

Criterion 3: Research and Development Must Focus on Improvement

Scientists and engineers have a wide range of experience with various proposed technologies of remediation. Obviously, we have no experience with terraforming deserts into jungles, and such proposals are as uncertain in their effects as proposals to seed the oceans with iron or inject aerosols into the stratosphere. On the other hand, there is some experience with the capture of carbon dioxide from power plants and the ambient air, as well as with various forms of sequestration. For instance, the injection of carbon dioxide into geological repositories has been done in the oil industry for years. None of this experience is at the scale needed, and for the most part it is in its early stages. However, there are technical cores to be worked from, offering some promise for technologies of carbon remediation.

Without a doubt, technologies of carbon remediation are today largely speculative and certainly expensive. At the same time, the technologies appear to have some promise, and a simple analysis suggests that cost estimates for some of these technologies are in the same ballpark as the costs of conventional mitigation that have been estimated by leading climate assessments. Carbon remediation certainly does not offer anything like a silver bullet to accumulating carbon dioxide emissions. Nevertheless, it meets the Sarewitz and Nelson criteria for a successful technological fix and shows enough promise that is should be considered an option worth pursuing as some part of a comprehensive response.

Debates Over Geoengineering Are Here to Stay

The allure of geoengineering is much like that of any proposed technological fix. If we can avoid costly or difficult social change through a technological intervention, why would we do anything else?

The analysis in this chapter suggests that—with one important exception—the various technologies of geoengineering are not technological fixes. Most of the technologies of geoengineering fail miserably with respect to the logical and straightforward criteria for successful technological fixes presented by Sarewitz and Nelson. This does not mean that continued research on all of these technologies is a waste of time or money. Over the very long term, perhaps such research will lead to one or more of these technologies meeting the technological-fix criteria. And even if they do not meet the criteria, there is value in conducting the research. What the analysis in this chapter suggests is that we should not be looking to geoengineering as a solution—or even part of a solution—to addressing the accumulation of carbon dioxide in the atmosphere, again, with one important exception.

That exception is carbon remediation via the direct removal of carbon dioxide from the atmosphere and its long-term storage. In the area of geoengineering there are more and less practical proposals. The most fanciful include terraforming giant deserts, which do not meet the criteria of a successful technological fix. On the other hand, technologies of carbon capture from power plants and the ambient air through biological, chemical, or geologic processes offer the most promise. If the costs of carbon remediation over the twenty-first century are indeed comparable to those costs of more conventional mitigation presented in major assessment reports such as by the IPCC and Lord Stern, and air capture avoids complexities such as the development and implementation of carbon markets, the need for changes to lifestyles and the behavior of billions of people, threats (perceived or real) to economic development, or sweeping technological innovation and diffusion, one might wonder why carbon remediation has not been front and center in the political debate on climate change, as have other forms of geoengineering. With prominent scientists and others—such as Wallace

Broecker, Jim Hansen, Al Gore, and Richard Branson—advocating the promise of the capture and storage of carbon dioxide, it is unlikely to remain out of sight for much longer.

Some have suggested that the mere discussion of geoengineering technologies introduces a "moral hazard" into the climate-change debate, suggesting that if emissions can be directly removed from the atmosphere, then it reduces the need for mitigation activities in the short term. The University of Michigan's Edward Parson suggests that we should be cautious about air capture, not for technical or economic reasons, but because it might work too well: "Analyses identifying better response options in the future reduce the political pressure for near-term efforts, by providing well-founded supporting arguments for those who oppose near-term efforts to any degree and for any reason."[43] Such perspectives reflect ideological differences highlighted by Alvin Weinberg in the 1960s between techno fixers and those wanting to change how people behave.

Environmental policies have always reflected a mix of social and technical responses to problems. The challenge of reducing emissions of chlorofluorocarbons, which damage the ozone layer, was dealt with through both the development of technological substitutes as well as regulatory incentives. Reduction of pollution from power plants that leads to "acid rain" was motivated by societal innovations in the creation of new markets for pollution permits. Yet a counterexample is the challenge of removing arsenic in drinking water, which is accomplished through the technology of water filtration and not by the creation and implementation of "arsenic markets" or "arsenic taxes." Where technological solutions contribute effectively to environmental goals, they have typically been implemented. Climate management does not meet the criteria for a successful technological fix, and it has considerable unknown risks. Carbon remediation shows greater promise. So long as scientists and policy makers frame climate policy in terms of stabilizing concentrations of atmospheric carbon dioxide, then we should expect geoengineering to be a key part of the international climate policy debate.

CHAPTER 6

How Climate Policy Went Off Course and the First Steps Back in the Right Direction

MUCH AS THERE IS a qualitative disagreement about what the response to climate change ought to be—such as whether it should be a comprehensive policy aimed at a large-scale restructuring of society or focused on technological policies aimed at addressing a discrete set of problems that society faces or aspects of both—there is an even more fundamental disagreement about what climate change itself is. This isn't just a question of skeptics versus the convinced. Believe it or not, the main scientific and policy institutions responsible for climate change in the international arena do not even agree on what the phrase "climate change" actually means. For example, if you take a look at the "Summary for Policy Makers" of the 2007 IPCC science working group, you'll find this, in its first footnote: "Climate change in IPCC usage refers to any change in climate over time, whether due to natural variability or as a result of human activity. This usage differs from that in the United Nations Framework Convention on Climate Change, where climate change refers to a change of climate that is attributed directly or indirectly to human activity that alters the composition of the global atmosphere and that is in addition to natural climate variability observed over comparable time periods."[1]

The policy community thus has a very narrow definition of climate change, which in essence refers to the effects of the emission of greenhouse gases due to human activity on climate, when those effects exceed the bounds of natural climate variability. By contrast, the IPCC defines climate change much more broadly; for them, it means a change in the statistics of climate over a period of thirty to fifty years (or longer) beyond natural variability, irrespective of the cause of the change. Those two competing definitions have led directly to pathological policies, biased and politicized science, and a climate policy architecture that is, at its core, scientifically and practically unsound. Understanding the consequences of having two different definitions of climate change in use—definitions that have coexisted in the climate debate for at least twenty years—lies at the heart of understanding why climate policy has been off course from the start, and the steps needed to get back on track.

Consider the following thought experiment. Let's start with the real world, in which the human consumption of fossil fuels leads to emissions of greenhouse gases, which lead to changes in the climate, which in turn result in undesirable effects on people and the environment. Let's call this Greenhouse World. Now imagine an alternative world. In this alternative world everything is exactly as it is in Greenhouse World, but with one important difference. In this world, instead of the human use of fossil fuels leading to changes in climate, the source of change is instead a small strengthening of the intensity of the Sun. Let's call this world Bright Sun World. The changes in climate and effects on people and the environment are identical in both Bright Sun World and Greenhouse World; the two worlds differ only in the source of the changes in climate.

In my classes on policy related to climate change, I often introduce this thought experiment and then ask the students to discuss how their policy recommendations might differ between Greenhouse World and Bright Sun World. Someone in every class starts out by saying that in Bright Sun World we wouldn't need any policy beyond business as usual because the source of change is natural, coming from the Sun. If it is natural, they argue, then there is no problem. Indeed, this response would be entirely appropriate under the logic of the Climate Convention (that is, the United Nations Framework Convention on Climate Change). In fact, the dramatic changes in climate in Bright Sun World would not even be classified as a climate change under the convention's narrow definition. In my class, this perspective typically is quickly challenged when someone else points out that we would still want to adopt policies to respond to the effects of a change in climate—regardless of which world you find yourself in; if you live in a floodplain, for instance, you would still be concerned about your vulnerability to damaging floods. Further discussions involve geoengineering, and questions about whether a decision on intervention depends upon the source of the changes in climate.

The thought experiment begins to hint at the problems that might arise by defining climate change solely in terms of the effects of human changes to the composition of the atmosphere, specifically referring to long-lived greenhouse gases. This definition and corresponding narrow focus would make sense if long-lived greenhouse gases were the sole source of changes to the climate, and if societal and environmental impacts of climate were dominated by this effect. However, in the real world, neither is true. For instance, as discussed in Chapter 1, climate changes resulting from human-caused influences on the climate system other than those that affect the chemistry of the atmosphere—such as from particulates like black soot or land-use effects on climate—are excluded under the Climate Convention.[2] The IPCC adopts a broader definition of "climate change" that is more scientifically accurate.[3] Claims that climate policy should be based on the work of the IPCC typically fail to recognize that the policy community has rejected the most fundamental statement of the IPCC on the issue—the very definition of "climate change."

Pathological Policies

The two-definition problem has been recognized for some time. For instance, John Zillman, an Australian scientist and an active participant in the IPCC, wrote in 1997, "When the IPCC says 'climate has changed over the past century,' it is simply saying the climate now is not the same as it was a century ago (whatever the cause), whereas the [Climate Convention] listener will reasonably interpret such a statement as the scientific community affirming that human influence has changed climate over the past century." This, he argued, can only confuse the public.[4] Indeed, such confusion is rampant in public and media discussions of climate change.

One consequence of the Climate Convention's narrow definition of climate change is that it necessarily implies that all of climate policy can be reduced to greenhouse gas policies, and, because the convention represents all greenhouse gases in terms of carbon dioxide equivalencies, climate policy is necessarily carbon dioxide policy.[5] The logic of this approach is seductively simple and elegant. The problem can be viewed as a traditional pollution challenge: human emissions of greenhouse gases will lead to changes in the global climate, and these changes will have significant negative impacts on environment and society. The logic of the response is equally straightforward: avoiding the predicted downside of climate change simply means stopping the actions—emissions—leading to trouble. There has been a rich set of invocations of this logic throughout the climate debate over the past twenty years, such as in President Bill Clinton's 2000 State of the Union address: "If we fail to reduce the emission of greenhouse gases, deadly heat waves and droughts will become more frequent, coastal areas will flood, and economies will be disrupted. That is going to happen," he said, "unless we act."[6] The implication is, of course, that if we act on greenhouse gases, those bad things will not happen. In September 2009 President Barack Obama invoked the same logic:

> Rising sea levels threaten every coastline. More powerful storms and floods threaten every continent. More frequent droughts and crop failures breed hunger and conflict in places where hunger and conflict already thrive. On shrinking islands, families are already being forced to flee their homes as climate refugees. The security and stability of each nation and all peoples—our prosperity, our health, and our safety—are in jeopardy. And the time we have to reverse this tide is running out.
>
> And yet, we can reverse it. John F. Kennedy once observed that "Our problems are man-made, therefore they may be solved by man." It is true that for too many years, mankind has been slow to respond or even recognize the magnitude of the climate threat. It is true of my own country, as well. We recognize that. But this is a new day. It is a new era.[7]

From this perspective carbon dioxide policy is like a big knob on a control panel through which global policy makers can modulate the future behavior of the climate system in a manner that modulates sea-level rise, disaster impacts, and famines.[8] This metaphor was made explicit as long ago as 1990 when Senator Ernest Hollings, a Democrat from South Carolina, likened the Earth to a car, noting, "When we have a car problem, we take the car to a repair shop or fix it ourselves using the operator's manual. For the global environment, however, there are no mechanics or manuals." He continued, arguing that society must "obtain the knowledge we need to train the mechanics and write the manual before this global machinery is irreversibly damaged." The metaphor was also invoked by Richard Alley, a climate scientist at Penn State University, who titled his featured lecture at the 2009 American Geophysical Union, "The Biggest Control Knob: Carbon Dioxide in Earth's History."[9] And the metaphor is reinforced by the home page of the European Commission, which represents climate policy in terms of a giant control knob on the Earth's surface (Figure 6.1), stating, "Even small changes in our daily behaviour can help prevent greenhouse gas emissions without affecting our quality of life."

The control-panel metaphor is also implicit in calls to tune atmospheric concentrations of greenhouse gases in such a way as to maintain a specific temperature increase, such as 2 degrees Celsius.[10] Policies to influence emissions are widely considered to be the primary policy tool to control the "global machinery" and thereby avoid future climate impacts. However, as we've seen, some have begun to suggest that various geoengineering strategies offer an alternative set of knobs on the climate-control panel.

The focus in the Climate Convention on only those climate changes that result from anthropogenic greenhouse gas emissions means that a prerequisite for action, politically if not practically, is the ability to identify climate changes related to the greenhouse gas emissions. In the jargon of the climate community, identification of climate changes and their causes is called detection and attribution. The need for science to detect and attribute climate change is codified in the Climate Convention's Article 2, which states that the convention's ultimate objective is "stabilization of

FIGURE 6.1 European Commission's "You Control Climate Change" campaign. Source: Copyright © European Communities.

greenhouse gas concentrations in the atmosphere at a level that would prevent dangerous atmospheric interference with the climate system."[11] Under the Climate Convention, without such detection and attribution there would be no reason to act, as there would be no evidence of climate change under its narrow definition, and certainly no danger.

The notion of "dangerous interference" is the policy expression of the narrow definition of climate change. The implementation of the climate convention depends upon determining some threshold above which climate change becomes dangerous and detecting that change and attributing it to greenhouse forcing. If climate change is not detected, or is not attributed to greenhouse gas forcing, then the Climate Convention has no formal basis for justifying action. While this approach may have created a clear separation between the climate issue and broader development activities of the United Nations, it has also influenced the dynamics of climate policy.

One consequence is that preexisting interests get mapped onto the notion of "dangerous." If the threshold of dangerous interference is subject to interpretation, then it becomes possible (and convenient) for partisans of various points of view to equate the threshold with political positions that have been determined through other means. For example, the administration of George W. Bush, quite vocal in its opposition to policies under the Climate Convention, claimed that "no one can say with any certainty what constitutes a dangerous level of warming, and therefore what level must be avoided." One scholar observes that reactions to the Climate Convention's Kyoto Protocol, like those to a Rorschach test, "generally reveal more about the speaker than about the protocol."[12] Not only does the notion of "dangerous interference" compel science to serve as the arbiter of what ultimately are political considerations that science cannot resolve, but it is inconsistent with how the climate in all of its complexity actually affects society and the environment, which themselves are undergoing changes for a range of reasons that also contribute to impacts of concern.[13]

Article 2 is an obstacle to effective action on climate change because of its focus on the notions of both "dangerous" and "interference." The phrase "dangerous interference" suggests that a threshold exists that separates a "dangerous" interference from one that is "not dangerous." But drawing such a line is not straightforward. Reality is much more complex. Society and the environment undergo constant and dramatic change as a result of human activities. People build on exposed coastlines and floodplains. Development, demographics, wealth, policies, and political leadership change and evolve over time. These factors and many more contribute to the vulnerability of populations to the impacts of climate-related phenomena. Different levels of vulnerability help to explain, for example, why a tropical cyclone that makes landfall in the United States has profoundly different impacts than a similar storm that makes landfall in Central America. Consequently, the degree to which climate is "dangerous" differs around the world and further depends upon how different communities value security and risk. The IPCC asserts that defining "dangerous interference" (as found in Article 2 of the FCCC) necessitates "value judgments determined through socio-political processes, taking into account considerations such as development, equity, and sustainability, as well as uncertainties and risk."[14] In a world where, for many communities, climate is already quite "dangerous," identifying a threshold becomes a matter of judgment, subject to differing perspectives and interests.[15] But "dangerous" also is variable in an objective sense, precisely because vulnerability varies with levels and patterns of development and other societal factors.

But not only is the notion of what is "dangerous" problematic, so too is the notion of "interference." This is the case for two reasons. First, because the adverse effects of climate are the consequence of human and climate (and other environmental) variables, there are many reasons a particular community or ecosystem may experience adverse climatic impacts under conditions of no climate change, human caused or otherwise. For example, a historic flood in an unoccupied floodplain may be a noteworthy curiosity, but a similar flood in a vastly populated floodplain is a disaster. The development of the floodplain could be the change that results in the phenomenon becoming dangerous; thus, the interference that leads to adverse impacts results from human occupancy of the floodplain. Under the Climate Convention, any such change would not be cause for action, even though adverse effects may still result. Climate occurs in a context of dramatic and rapid societal changes that affect not only society itself but the environment society inhabits. In many contexts, the most important "interference" does not involve influences on the climate but instead involves how societal changes increase the exposure of society and the environment to powerful events—like hurricanes or floods—that already exist.[16] Chapter 7 will explore this issue in some detail.

A second challenge in documenting "interference" has to do with the nature of the global Earth system itself. Climate changes at all timescales and for many reasons, not all of which are fully understood or quantified. Consider, for example, abrupt climate change that might take place over a relatively short time period, such as a decade or less. A review paper in *Science* that I coauthored observes that "such abrupt changes could have natural causes, or could be triggered by humans and be among the 'dangerous anthropogenic interferences' referred to in the [Climate Convention]. Thus, abrupt climate change is relevant to, but broader than, the [Climate Convention] and consequently requires a broader scientific and policy foundation."[17] Like most issues associated with climate change, greenhouse gas emissions are only a subset of reasons for concern, yet the climate convention focuses on this issue to the exclusion of the broader considerations. In an important respect, the phrase "climate change" is redundant and defining it exclusively with the effects of greenhouse gas emissions is scientifically incomplete, if not misleading.

Consider another example. A group of researchers, including my father, has suggested that changes in regional land-use patterns have potential to alter regional and global climate, yet the Climate Convention concerns itself only with human society's emissions of greenhouse gases. The Climate Convention, they argue, needs to start paying attention: "Mitigation strategies that give credits or debits for changing the flux of carbon dioxide to the atmosphere but do not simultaneously acknowledge the importance of changes in the albedo or in the flows of energy within the Earth system might lead to land management decisions that do not produce the intended climatic results."[18] In other words, these researchers raise the possibility that efforts to extract carbon dioxide from the atmosphere and store it in vegetation may have the perverse effect of changing the energy balance of the earth system, resulting in an additional source of human disruption of the climate system, the exact opposite of the intentions for biological sequestration. The definition of climate change under the Climate Convention does not even formally recognize land-use effects on climate, as it is only concerned with the effects of greenhouse gases. Thus, in a perverse sense, the Climate Convention could in principle succeed in limiting the effects of greenhouse gases on the climate while actually creating other undesirable changes in climate in the process.

Preventing interference in the climate system by focusing only on greenhouse gas forcing makes sense from a scientific perspective only if other potential natural and human-caused changes in the climate system are by comparison insignificant. This assumption appears to be the perspective of the Climate Convention, because in 1996 its leaders argued that its narrow definition of climate change did not differ significantly from that of the IPCC because "in many instances the two uses will in effect be the same, and this is particularly true for projections of climate change over the next century."[19] In other words, carbon dioxide is all that really matters. If this assumption about the basic science of the global earth system is incorrect, then the Climate Convention has set the stage for significant problems in its implementation.[20] At the very minimum, we should not equate climate policy with carbon dioxide (or even greenhouse gas) policy—climate policy is far more complex.

In short, the idea that science can detect and attribute interference in the climate system related only to greenhouse gas forcing is problematic; climate changes on all timescales because of a range of both natural and human factors. And even if science could detect and attribute climate change, such changes occur in a world in which climate can already be dangerous to varying degrees, because of both differing perceptions of what is or is not dangerous and decisions that affect socioeconomic conditions that, in turn, affect vulnerability, and hence the degree of danger. Because of the illogic of Article 2 of the Climate Convention, considerably more attention has been paid not only by researchers but also by political advocates to the details of detection and attribution than to providing decision makers with useful knowledge that might help them to improve energy policies and reduce vulnerabilities to the climate.[21]

Consequences of Misdefining Climate Change for Policy

There could be a number of reasons the IPCC and the Climate Convention have different definitions of the phrase "climate change." It could have occurred for intellectual reasons, such as the assumption that the climate system is otherwise stationary absent an anthropogenic greenhouse gas forcing; for pragmatic reasons, such as the prior existence of international efforts focused on development and natural disasters; or for political reasons, such as the fact that a focus on greenhouse gases places the problem in the domain of energy policy. Whatever the underlying reasons for the different definitions, it is clear that it has had a pathological impact on the policy and politics of climate change. Specifically, the narrow definition of climate change in the Climate Convention leads specifically to bias against adaptation and more generally to the politicization of science.

A Bias Against Adaptation

For decades, the options available to deal with climate change have been clear. We can act to mitigate the future impacts of climate change by addressing the factors that cause changes in climate. And we can adapt to changes in climate by addressing the factors that influence societal and environmental vulnerabilities to the effects of the climate. Mitigation policies focus on either controlling the emissions of greenhouse gases or capturing and sequestering those emissions. Adaptation policies focus on taking steps to make social or environmental systems more resilient to the effects of climate. Effective climate policy will necessarily require a combination of mitigation and adaptation policies.[22] However, climate policy has for the past decade reflected a bias against adaptation in large part due to the differing definitions of climate change.

The bias against adaptation is reflected in the IPCC's inconsistent attitude toward its own definition of climate change. Its working group on science prefers (and indeed developed) the broad IPCC definition, including both anthropogenic and natural change. The working group on mitigation prefers the narrow definition of the Climate Convention in order to be more directly relevant to the discussions under the Climate Convention, and the working group on impacts, adaptation, and vulnerability uses both definitions at different points in its analysis.[23] One result of this inconsistency is an implicit bias against adaptation policies in the IPCC reports and, by extension, in policy discussions. As the limitations of mitigation-only approaches emerge and policy making necessarily turns toward adaptation, addressing this bias becomes increasingly important.

Under the Climate Convention's definition, adaptation refers only to actions in response to climate changes attributed to greenhouse gas emissions. Absent the increasing greenhouse gases, climate—by definition—would not change, so adaptive measures would be unnecessary. This means that under the narrow definition, adaptation can have only costs; the measures would be necessary only because of the changes in climate that result from greenhouse gas emissions. That is, there can be no benefits—only costs incurred to maintain the status quo ante. This exclusion of benefits may seem like a peculiarity of accounting, but it has practical consequences. One IPCC report used the narrow definition to discuss climate policy alternatives in exactly this way, affecting how policy makers perceive alternative courses of action: it discussed mitigation policies in terms of both costs and benefits, but adaptation policies only in terms of their costs.[24] If a course of action has costs but no benefits, it will not fare well in policy evaluation! This stacks the deck against adaptation policies and ensures that mitigation will look better.

The bias against adaptation is particularly unfortunate not only because the world is already committed to some degree of climate change (due to greenhouse gases, but also other factors) but also because many communities around the world are maladapted to the current climate. Many, if not most, adaptive measures would make sense even if there were no greenhouse gas–related climate change.[25] The narrow definition of climate change provides little justification for efforts to reduce societal or ecological vulnerability to climate variability and change beyond those impacts caused by greenhouse gases. For instance, in 2004, the International Institute for Sustainable Development reported on concerns raised by poor countries in the climate negotiations that in order to receive funding they had to show the incremental damage caused by greenhouse gas–driven climate change. Simply experiencing climate impacts was not enough. Their inability to show this incremental effect kept international adaptation funding perversely out of reach.[26] From the perspective of the broader IPCC definition of climate change, adaptation policies also have benefits to the extent that they lead to greater resilience of communities and ecosystems to climate change, variability, and particular weather phenomena. Those who work under the Climate Convention are aware of these problems and try to work around them; however, work-arounds do not address the fundamental issues.

The restricted perspective of the narrow definition makes adaptation and mitigation seem to be opposing strategies rather than complements and creates an incentive to recommend adaptive responses only to the extent that proposed mitigation strategies cannot prevent changes in climate. For many in the climate debate, adaptation represents the costs of failed mitigation and thus represents an outcome to be avoided, rather than opportunities to make communities and ecosystems more resilient and robust. From the perspective of adaptation, the approach of the Climate Convention serves as a set of blinders, directing attention away from adaptation measures that make sense under any scenario of future climate change. As nations around the world necessarily move toward a greater emphasis on adaptation in the face of the obvious limitations of mitigation-only policies, reconciling the different definitions of climate change becomes more important.

Politicization of Climate Science

A February 2003 article in *The Guardian* related details of the climate policy debate in Russia and revealed the absurdity of the dominant approach to climate policy. The article reports that several Russian scientists "believe global warming might pep up cold regions and allow more grain and potatoes to be grown, making the country wealthier. They argue that from the Russian perspective nothing needs to be done to stop climate change." In other words, they believe that not only will climate change not result in "dangerous interference," but it will result in what might be called "beneficial interference." As a result, "To try to counter establishment scientists who believe climate change could be good for Russia, a report on how the country will suffer will be circulated in the coming weeks." Science was thus enlisted not only to show that human activities affect the climate but also to show that resulting changes will become dangerous in the political sense of the term. Such arguments had been made before. For example, Klaus Topfer, executive director of the United Nations Environment Program, protested in 2001: "There are no winners, only losers, in the climate change scenario. Now is the time to act collectively and decisively."[27]

Why does this matter? The terms of the Climate Convention force political combatants to assert certainty about the climate future (dangerous or not?) when in reality uncertainty may be irreducible. At the time, such certainty was deemed to be necessary to promote or campaign against ratification of the Kyoto Protocol. For example, Paul Jeffries, head of environmental policy at Britain's Royal Society for the Protection of Birds, explained why it was so critically important to resolve scientific questions about climate impacts on Russia: "Russia's ratification [of the protocol] is vitally important. If she doesn't go ahead, years of hard-won agreements will be placed in jeopardy, and meanwhile the climate continues to change."[28] Thus, any scientific result that suggests that Russia might benefit from climate change, or even uncertainty about future impacts, could be used to argue that climate change would not be "dangerous" for Russia and thus be used to counter the implications of Article 2 of the Climate Convention that seeks to prevent dangerous interference. When political battles are waged through science, uncertainty is often underplayed on all sides of the debate. Science that suggested large climatic impacts on Russia was used to support arguments for Russia's participation in the protocol. In this manner, the science of climate change becomes irrevocably politicized, as partisans on either side of the debate selectively array bits of science that best support their position. The Climate Convention sets forth the notion of "dangerous" as a political dividing line. The reality is that science cannot definitively tell us what is dangerous or not, as the future is always to some degree uncertain, but so too is how we value different future outcomes.

The narrow definition of climate change provides a political motivation to produce or spin science that shows or dispels "dangerous interference." The Climate Convention makes it difficult to consider, much less enact, policies that do not depend upon certainty in future outcomes or are robust with respect to the climate future, irrespective of the source of change.[29] There is no room under Article 2 for irreducible uncertainties or fundamental ignorance about the climate future. Conversely, the IPCC notes that climate change requires "decision making under uncertainty."[30]

These arguments reflect the prescience of an argument presented by Mickey Glantz in 1995: "If winners and losers are identified with some degree of reliability, the potential for unified action against the global warming may be reduced. Winners will not necessarily want to relinquish any portion of their benefits to losers in order to mitigate the impacts of their losses. " Glantz further notes that ignoring the issue of winners and losers can be problematic as well: "While scientists and policymakers formally discuss only losses associated with a global warming, others may perceive that there will be positive benefits as well. . . . This could sharply reduce the credibility of the proponents for taking action, lessening the chances for *any* response, preventive, mitigative, or adaptive."[31]

Not only does the Climate Convention create a bias against adaptation; it also encourages claims to certainty about the future that inevitably lead to a politicization of the science of climate change.[32] An approach that is more consistent with the realities of science and needs of decision makers would begin with a framing commensurate with these realities. Under the Climate Convention climate change is viewed as a single problem, when in fact it is many. The Climate Convention further encourages arguments that emphasize certainties (or known uncertainties) and thus de-emphasizes decision making under uncertainty and ignorance. The Climate Convention asks science to provide answers that simply are not forthcoming.

Getting Climate Policy Back on Course

There is no doubt that the Climate Convention and its Kyoto Protocol represent tremendous diplomatic accomplishments. Negotiations have raised awareness of countries around the world to the importance of the climate and focused them on shared objectives. But at the same time, this approach does not get at the core societal and environmental problems of climate change. One essential factor in improving the Climate Convention is to consider how it might be structured in the context of the broader definition of climate that allows for a clearer distinction of carbon dioxide policy and broader issues of climate policy. Such an approach offers no panacea to dealing with the challenges of climate change, but at a minimum it offers a radical reframing of the issue that may open up discussion of paths not seen and options not previously considered.[33]

In terms of climate policy, such a reframing would mean a transition from a focus on "dangerous human interference with the climate system" to a much narrower concentration on the individual factors that contribute to climate change. One such focus would be on accumulating greenhouse gases, specifically carbon dioxide. Trying to reach a political consensus on how much carbon dioxide in the atmosphere is "dangerous" overlooks the fact that stabilizing concentrations at any level means transforming the global energy system—including diversifying the energy supply while pursuing energy security and access—while considering potential backstop technologies.

A similar, but distinct, focus might be placed on other factors deemed to be important contributors to climate change under separate policy instruments. The Montreal Protocol on the depletion of the ozone layer has been suggested as a policy vehicle to address several other greenhouse gases. Methane and nitrous oxide, two other important greenhouse gases, might be considered under separate policy instruments. More broadly, policies might be developed to focus on black carbon, land-use effects on climate, and other human influences.

A third concern must be adaptation, defined as better preparing for the impacts and opportunities related to the climate, recognizing that the word "climate" by definition includes the notion of variability and change. That would bring adaptation much closer to traditional notions of development and would emphasize those policies that actually serve to diminish environmental degradation and improve human lives.[34] Building resilience to human-caused climate change is a notable side benefit to implementing adaptation policies that make sense on their own merits.

The basic principle of a more mature international adaptation policy is that the climate "winners" of the world would bear some responsibility for the climate "losers" of the world. Of course, the international community has for many years discussed disaster relief, debt forgiveness, and development assistance. A new approach to climate policy would distinguish such issues more clearly from the issues associated with carbon dioxide. When, at some point in the future, the distribution of climate winners and losers changes, then the relative roles and responsibilities of nations would change accordingly. In this manner, a climate adaptation policy could lead to immediate, demonstrable results. There would be no need to rely on documentation of changes in climate or attribution to a human cause as the basis for action. The impacts of climate are painfully apparent and are with us today, not in some hypothetical future. Actual decisions about how best to reduce societal and environmental vulnerability to the climate would be made in local contexts based on assessments of costs, benefits, and risks, as well as local values about the climate as a resource and a threat.

Arguably, climate policy has already begun to move in the directions suggested here, though haltingly and slowly, motivated more by successive policy failures than any coherent vision for an alternative approach. A reworked approach to climate policy might empower decision makers to follow the old environmental adage, "Think globally, act locally," rather than pursuing mistaken attempts to think globally and act globally.

An approach that decouples carbon dioxide policy from other aspects of climate policy would create a more productive role for the scientific community as well. There would remain need for periodic snapshots of the state of the science, as currently done via the IPCC. But the sorts of questions to be addressed would change dramatically, such as being directly driven by the needs of policy makers facing specific challenges at local and regional scales of governance. With respect to carbon dioxide, there would be much more attention paid to innovation in energy technologies and energy systems. There would be a decreased emphasis on research that seeks to attribute or predict changes in climate over century-long timescales, because policy action would no longer be dependent upon a presumption of accurate predictions that allow judgments of dangerous interference. There would be instead an increased emphasis on research that seeks to understand the interactions of climate, society, and environment in ways that lead to vulnerabilities (as well as opportunities) in local and regional contexts, rather than on global scales. Research would focus more on providing information useful for addressing the problems of today—such as malaria and extreme events—that we know will also be the problems of tomorrow. An emphasis on policy-centered research under conditions of irreducible uncertainty and ignorance would help decision makers to evaluate what sorts of actions work to reduce vulnerabilities and which ones do not. Science would thus place itself in the role of being a tool for policy action rather than primarily a tool for political advocacy. Science has been moving in this direction, but too slowly, and it has been held back by the focus of the Climate Convention.

Uncertainty—whether in predictions of the future climate or attributing specific climate events to human emissions of greenhouse gases—is not going to disappear in the foreseeable future. Kevin Trenberth, a climate scientist at NCAR, argues that uncertainties are in fact going to get larger in the near term: "The uncertainty in [the 2013 IPCC] climate predictions and projections will be much greater than in previous IPCC reports" due to the employment of more complex models.[35] That uncertainty will inevitably fuel continued public debate.[36] And even if uncertainties about the future were to be reduced, as Glantz has noted there is no reason to believe that would make the politics any easier. On the one hand, this seems to suggest that scientists will benefit as each side of the debate demands greater certainty and looks to science to provide that certainty.[37] But on the other hand, more research could very easily lead to greater uncertainties, and thus there exists a real possibility that the scientific community could suffer a backlash of public criticism that affects not only their role in the climate issue but also public support for climate science more generally. Climate science offers the promise of great benefits to humanity; it is incumbent upon the scientific community to reshape the current debate in ways that enhance the contributions of research to worthwhile objectives.

A critical first step in reshaping the current debate is to highlight the pervasive consequences of the narrow definition of climate change used under the Climate Convention and to consider how policy might be more effective if redesigned under a broader perspective. With a reframed policy that decouples carbon dioxide policy from other aspects of climate policy, the community of scientists, advocates, and diplomats might find the surprising result that they will not only see multiple paths to reduce human and environmental vulnerability to the climate but also create more effective possibilities to achieve in practice the goal of accelerated decarbonization of the global economy. I'll return to these themes and offer more specific guidance on a reformulated approach to climate policy in Chapter 9. But first, I turn attention to some of the specifics of adaptation in the context of extreme events.

CHAPTER 7

Disasters, Death, and Destruction

ADAPTATION POLICIES make sense regardless of uncertainties about climate science or the potential physical and economic impacts of emissions of greenhouse gases. Quite simply, this is because the root causes of the disasters, death, and destruction that so often characterize the relationship of humans and their global environment don't need climate change (anthropogenic or otherwise) to exist. For a disaster to occur, only two conditions must coexist: an extreme event and a vulnerable society. The chapter will explain trends in disasters over the past century, with a focus on recommending policy approaches that will yield robust results whatever the future of the climate has in store for us. In particular it will focus on hurricanes and floods, which are responsible for the vast majority of property damage related to weather events around the world.

In 2006 I organized a major international workshop on disasters and climate change with Peter Höppe, who runs the Geo-Risk division for Munich Reinsurance, a large global company headquartered in Germany that provides insurance to insurance companies. The purpose of our workshop, held in Hohenkammer, Germany, just outside of Munich, was to examine trends in disaster losses and to see if we could explain why losses had been increasing dramatically in recent decades and, in particular, to see if we could detect a signal of the effects of accumulating atmospheric greenhouse gases in the rising toll of disasters. The workshop originated in a debate that Peter and I had been having off and on for a while: he thought (and still does) that the data, which indicate a dramatic increase in disaster losses, contained a signal of climate change due to greenhouse gas emissions, whereas I thought that the evidence did not yet support such a conclusion. Our workshop rigorously examined what the state of the science would allow us to say and not say on this topic, which often is at the center of advocacy and debate over climate change.

In light of the issues involved with conflicting definitions of climate change, as discussed in the previous chapter, we were extremely careful to distinguish climate change as used by the IPCC, meaning a change regardless of cause, from the narrow definition of climate change as resulting from greenhouse gas emissions. Before the meeting Peter and I had planned on arriving at what we called a "consensus dissensus," that is, we would agree collectively at the meeting on where we were unable to reach a group consensus. But much to our surprise and delight, all thirty-two people at the workshop—experts from academia, the private sector, and advocacy groups—reached a consensus on twenty statements on disasters and climate change. (These statements can be found in the text box on page 163.)

At the end of the workshop, we arrived at several conclusions about the influence of climate change resulting from an increase in greenhouse gases on disasters. They include:

1. Analyses of long-term records of disaster losses indicate that societal change and economic development are the principal factors responsible for the documented increasing losses to date.
2. Because of issues related to data quality, the stochastic nature of extreme-event impacts, length of time series, and various societal factors present in the disaster-loss record, it is still not possible to determine the portion of the increase in damages that might be attributed to climate change due to greenhouse gas emissions.
3. In the near future the quantitative link (attribution) of trends in storm and flood losses to climate changes related to greenhouse gas emissions is unlikely to be answered unequivocally.

Consensus (Unanimous) Statements of the Hohenkammer Workshop Participants

1. Climate change is real and has a significant human component related to greenhouse gases.
2. Direct economic losses of global disasters have increased in recent decades with particularly large increases since the 1980s.
3. The increases in disaster losses primarily result from weather-related events, in particular storms and floods.
4. Climate change and variability are factors that influence trends in disasters.
5. Although there are peer-reviewed papers indicating trends in storms and floods, there is still scientific debate over the attribution to anthropogenic climate change or natural climate variability. There is also concern over geophysical data quality.
6. The IPCC's 2001 report did not achieve detection and attribution of trends in extreme events at the global level.
7. High-quality, long-term disaster-loss records exist, some of which are suitable for research purposes, such as to identify the effects of climate or climate change on the loss records.
8. Analyses of long-term records of disaster losses indicate that societal change and economic development are the principal factors responsible for the documented increasing losses to date.
9. The vulnerability of communities to natural disasters is determined by their economic development and other social characteristics.
10. There is evidence that changing patterns of extreme events are drivers for recent increases in global losses.
11. Because of issues related to data quality, the stochastic nature of extreme-event impacts, length of time series, and various societal factors present in the disaster-loss record, it is still not possible to determine the portion of the increase in damages that might be attributed to climate change due to greenhouse gas emissions.
12. For future decades the IPCC 2001 expects increases in the occurrence and intensity of some extreme events as a result of anthropogenic climate change. Such increases will further increase losses in the absence of disaster-reduction measures.
13. In the near future the quantitative link (attribution) of trends in storm and flood losses to climate changes related to greenhouse gas emissions is unlikely to be answered unequivocally.

(continues)

Policy Implications Identified by the Workshop Participants

14. Adaptation to extreme weather events should play a central role in reducing societal vulnerabilities to climate and climate change.
15. Mitigation of greenhouse gas emissions should also play a central role in response to anthropogenic climate change, though it does not have an effect for several decades on the hazard risk.
16. We recommend further research on different combinations of adaptation and mitigation policies.
17. We recommend the creation of an open-source disaster database according to agreed-upon standards.
18. In addition to fundamental research on climate, research priorities should consider the needs of decision makers in areas related to both adaptation and mitigation.
19. For improved understanding of loss trends, there is a need to continue to collect and improve long-term and homogenous data sets related to both climate parameters and disaster losses.
20. The community needs to agree upon peer-reviewed procedures for normalizing economic loss data.

Since that workshop in 2006, the evidence in support of these conclusions has, if anything, gotten much stronger. This chapter will discuss extreme events and their impacts, with a focus on understanding why it is so difficult to see a signal of greenhouse gas–driven climate change in the disaster record. The difficulty in detecting a signal provides an important clue that explains why adaptation policies make sense regardless of the exact course of climate change in coming years and decades.

This chapter will also discuss an uncomfortable aspect of the climate debate: the systematic misrepresentation of the science of disasters and climate change not just in public debates, but—arguably more problematically—in the leading scientific assessments produced to inform policy. The misrepresentation is related to the political dynamics encouraged by the narrow definition of climate change discussed in the previous chapter. The lessons from the case of disasters can help to inform adaptation policies in response to other human and ecological impacts associated with climate, such as disease, sea-level rise, and impacts on natural systems.

The policy implications of the arguments presented in this chapter are as clear as they are uncomfortable. Reducing the losses associated with disasters that will happen in coming decades is almost entirely a matter of effective adaptation policies. Using disasters to advocate for mitigation policies is misguided at best and misleading at worst. As I have argued in earlier chapters, there are indeed very good reasons to advocate for a decarbonization of the global economy. Nevertheless, as I'll show, decarbonization policies are not particularly appropriate means of modulating the impacts of future disasters. Unfortunately, the temptation to justify energy policies based on imagery of disasters is too great for some to resist, despite the tenuous scientific basis for making a connection between the two. Succumbing to this temptation has led to some of the most egregious errors in leading scientific assessments of climate change. But before we get to that issue, we must first understand the relationship of climate and societal change.

At our 2006 workshop we focused on understanding the long-term increase in disaster losses, which is often presented using an iconic figure from Munich Reinsurance (redrawn here as Figure 7.1). It shows that both total and insured losses from all major disasters (including earthquakes and other non-weather-related phenomena) have increased dramatically in recent decades.[1]

Figure 7.2 shows where we should look if we want to understand the reasons for the increasing losses. It shows the frequency of phenomena that lead to the total losses displayed in Figure 7.1. The darkest part of the bars show geological hazards, which are not directly related to climate. The dark-gray parts of the bar show phenomena such as temperature extremes, drought, and wildfire, which are certainly important but nonetheless are minor contributors to total global disaster losses. (That said, I will return to this subject later.). The white and light-gray parts of the bar represent floods and windstorms, respectively; these are the primary contributors to the global increase in disaster losses; if we can figure out why they're increasing, we'll have largely figured out what is driving the overall global trend of increasing disaster losses.

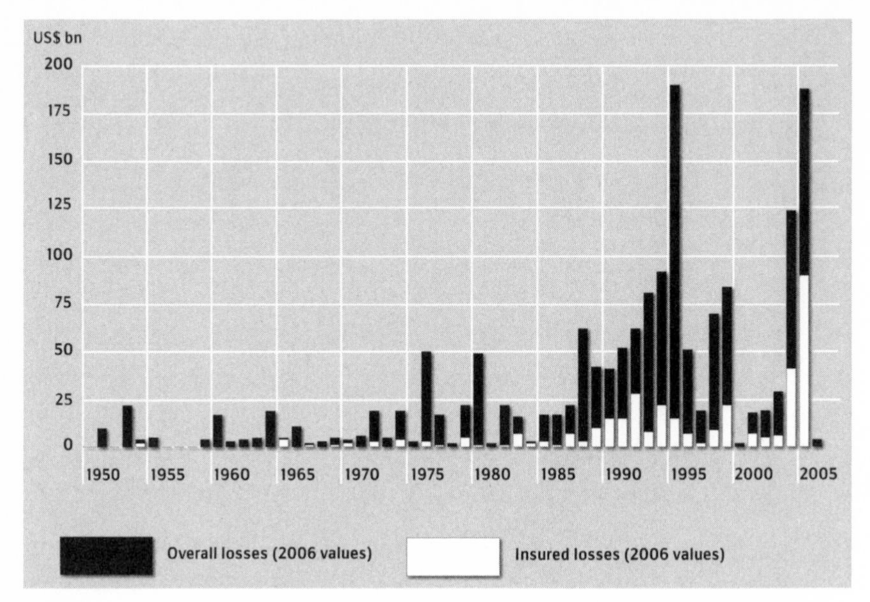

FIGURE 7.1 Total losses (insured and uninsured) due to disasters since 1950. Source: Munich Re.

Understanding Tropical Cyclone Disasters Over Time and Around the World

Hurricanes are what strong tropical cyclones are called in the North Atlantic and eastern Pacific oceans. Globally, there are about ninety hurricane-strength tropical cyclones each year, and they are a terrifying phenomenon. In 1991 a strong tropical cyclone making landfall in Bangladesh killed more than 140,000 people and displaced more than 10 million. In 2005 Hurricane Katrina caused more than $80 billion in damage, making it the most costly weather-related disaster ever.

Understandably, the tropical cyclone has become a sort of poster child for the climate debate, making appearances everywhere from the cover of Al Gore's most recent book to the opening video presented to delegates at the 2009 UN climate conference in Copenhagen. In other examples (see Figure 7.3), a leading environmental group used the image of a knob on a control panel as a metaphor to suggest that clean energy was "the solution" that could be used to determine if hurricane strength was bad, real bad, or really, really bad. In the 2004 presidential

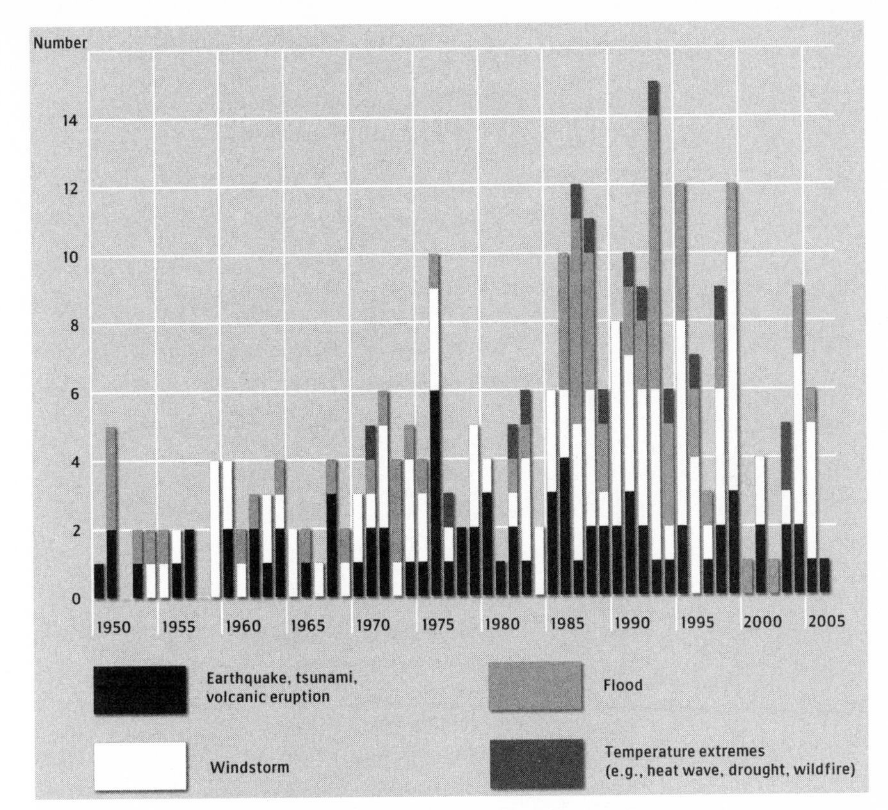

FIGURE 7.2 Incidence of disasters since 1950. Source: Munich Re.

election between George W. Bush and John Kerry a group called Scientists and Engineers for Change posted billboards in the state of Florida suggesting that the outcome of the election would influence hurricanes that threaten Florida. And, of course, the movie poster to *An Inconvenient Truth*, Al Gore's documentary on global warming, featured a tropical cyclone (from the Southern Hemisphere, oddly enough) emerging from the smokestack of a power plant—with the clear implication that shutting down the power plant would make the storm disappear.

One of the chief problems with this imagery is that the primary driver of increasing disaster losses around the world is not climate change, human caused or otherwise. It is development, which has led to more people and wealth in locations exposed to extreme events—not an increase in the frequency of extreme events themselves. The

FIGURE 7.3 Hurricanes and global warming in popular media and political ads.

following case study involves U.S. hurricanes, but it should be considered representative of what has been found everywhere that scholars have looked: societal factors have historically and will, for the foreseeable future, overwhelmingly drive increasing disaster losses, even assuming large changes in climate due to greenhouse gas emissions or other factors.[2] But to understand this argument requires getting into the data a bit.

In the mid-1990s the media, policy makers, the insurance industry, and some scientists were under the impression that greenhouse gas–driven climate change was leading to increasing disasters, even though at the time there was almost no research on this subject. For example, in early 1996 *Newsweek* had a cover story on extreme weather. The cover said, "Blizzards, Floods and Hurricanes: Blame Global Warming." At the time I was in the middle of a postdoctoral research assignment at NCAR, where I was working on a project focused on extreme weather events and their societal impacts, with a focus on hurricanes and floods. It was fairly obvious that the perception that changes in climate were responsible for increasing hurricane losses, in particular, was simply incorrect. Although 1992 was, to that point, the most economically disastrous hurricane season ever, research at the time showed that hurricane activity had been declining for the previous five decades. Furthermore, the period from 1991 through 1994 was the quietest four-year period in that entire span.[3] The most damage occurred during the quietest period of hurricane activity—that is exactly the sort of incongruity that academics like to explore a bit deeper.

To investigate this issue, I began a collaboration with Chris Landsea, who was working for a U.S. government research lab in Miami studying hurricanes (the collaboration continues today, and Chris is now at the National Hurricane Center). We decided to ask a simple question: what would the U.S. damage from hurricanes look like if every hurricane season of the past had occurred with present-day societal conditions?[4] Figure 7.4 shows why adjusting for societal change is so important. The figure shows pictures of Miami Beach in 1925 and in 2006. While Miami appears at exactly the same place on a map at these two different points in time, it was a very different place eighty years apart. By holding societal conditions constant—and asking what would be the impact of historical events with today's societal conditions—we could better see any climate signals present in the loss data. Of course, those looking for climate signals in hurricane activity would be better off looking first at climate data rather than loss data. Nevertheless, if there is a climate signal of some sort in the loss data, it should stand out after adjusting for societal changes.

FIGURE 7.4 Aerial views of Miami Beach. Sources: *1925 photo:* Florida Photographic Collection, Florida Department of State, State Library and Archives of Florida, http://www.floridamemory.com. *2006 photo:* Joel Gratz.

In our work on hurricane damages we have adjusted the data for various factors: inflation, population, more buildings, and personal "tangible wealth," which is best thought of as the amount of "stuff" that people have in their homes and businesses. Our most recent study came out in 2008 and updated losses through 2005.[5] Figure 7.5 extends the analysis through 2009 and shows U.S. hurricane losses adjusted only for inflation to 2010 values, and indicates a dramatic rise in the magnitude of large loss events as well as in their frequency, much like the Munich Reinsurance global figure.[6] However, once we adjust the data to account for the massive coastal development, as shown in Figure 7.6, the upward trend disappears, leaving a time series that exhibits no trend.

We have confidence in these loss estimates for three important reasons. First, hurricane landfalls in the United States have not increased in frequency or intensity over the period our data set covers.[7] In the absence of stronger or more frequent storms it is simply logical that there is no reason to expect to see any increases in the adjusted damage, and that is indeed what our analysis shows. The very close relationship

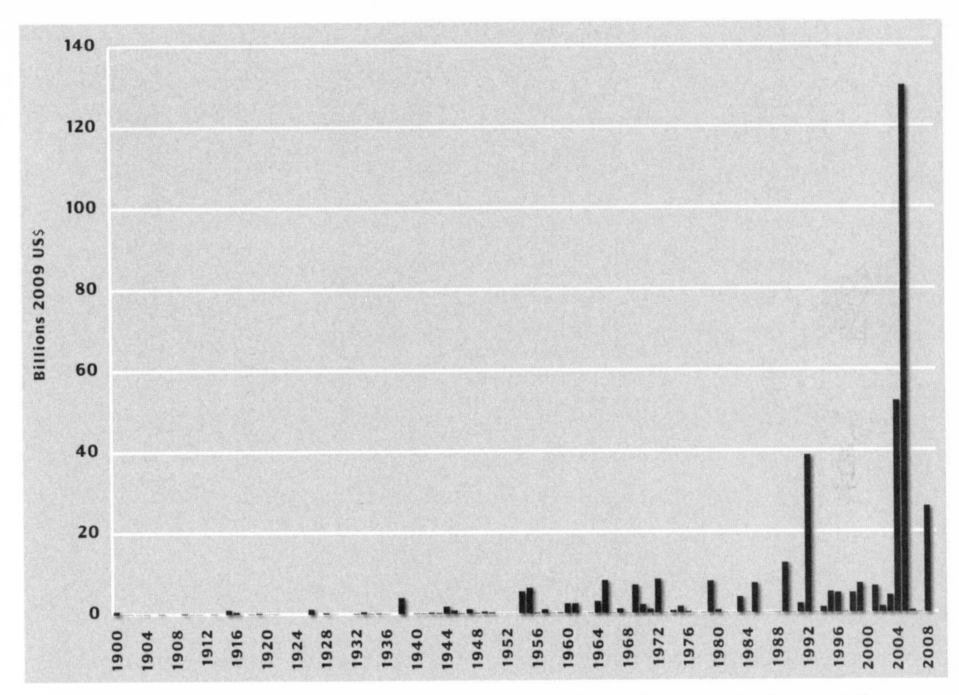

FIGURE 7.5 Inflation-adjusted U.S. hurricane losses, 1900–2009. Source: Author's calculations.

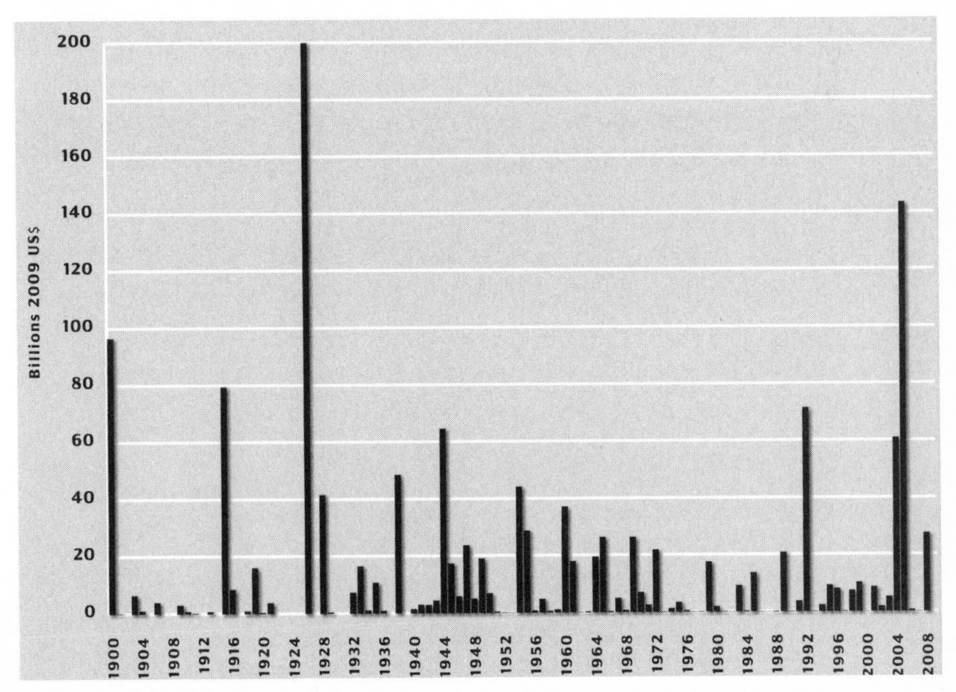

FIGURE 7.6 Normalized U.S. hurricane losses, 1900–2009. Source: Author's calculations.

between the physical statistics of hurricanes at landfall and our normalized loss results gives us good confidence that there are not other important factors left out of our normalizations, such as changes to building practices or codes. Second, we can identify signals of climate variability in the loss data. For instance, we can clearly see that years in which the Pacific Ocean is cool (called La Niña) have larger losses on average that do warm (called El Niño) or neutral years.[8] Finally, our work has been independently replicated by other scholars on several occasions using different approaches but arriving at confirming results.[9]

In the United States it is clear that hurricane losses have not increased any faster than would be expected in the context of a rapidly growing coastal population experiencing robust economic growth. So it would be misleading to attribute any of the increase in hurricane losses to climate change, human caused or otherwise. Scholars have done similar analyses of normalized tropical cyclone losses in Latin America, the Caribbean, Australia, China, and the Andhra Pradesh region in India. In each case they have found no trend in normalized losses. In fact, when one looks at tropical cyclone landfalls in continental regions around the world, no trends over the periods of record have been found in the United States, Mexico, Australia, China, and Southeast Asia, with only South Asia and northeastern Africa left to be documented. It should be obvious that with no trends in landfall occurrence, especially in the regions that contribute the most economic losses, that climate change (regardless of cause) cannot be a significant factor in explaining increasing disaster losses. A study published in early 2010 by a team of authors gathered by the World Meteorological Organization, which included scientists who had been at odds on the topic several years earlier, provided an unambiguous statement: "We cannot at this time conclusively identify anthropogenic signals in past tropical cyclone data."[10]

The bottom line is straightforward: it is misleading to suggest that the increasing toll of losses related to tropical cyclones has been influenced by accumulating greenhouse gases in the atmosphere. Looking to the future, patterns of development in exposed locations will continue to be the most important factor responsible for increasing losses in disasters, even assuming rather dramatic changes in tropical cyclones.[11]

Addressing those losses thus requires attention to where and how we develop, which in the climate issue falls under the category of adaptation. The story with floods is very much the same.

The Flood of Development

Although flood losses have not been studied as comprehensively as have tropical cyclones, where they have, very similar results have been found. For instance, research I have been involved in looked at the relationship of precipitation trends in the United States and flood damage. We found several interesting things in this research. First, the metric that scientists often use to define "intense" or "extreme" precipitation is often 50 millimeters (about 2 inches) of rain falling in one burst. However, this measure of rainfall is not at all well related to damages. In most places 2 inches of rain can fall without anyone really noticing, much less causing a major disaster. So even if occurrences of "intense rainfall" have increased, that metric does not imply a greater number of damaging floods.

The 2001 IPCC assessment discussed floods in some depth and reached a similar conclusion (the 2007 IPCC was largely silent on floods). Its authors noted "a widespread increase in heavy and extreme precipitation events in regions where total precipitation has increased," such as the middle and high latitudes of the Northern Hemisphere. But they cautioned that more heavy rain didn't necessarily mean more floods, and said, "Even if a trend is identified, it may be difficult to attribute it to global warming because of other changes that are continuing in a catchment." Not every analysis has even found increased rainfall. A 2005 review paper in the *Journal of Climate* by an international team of researchers concluded that statistically noisy data and uncertain models meant that "anthropogenic rainfall changes cannot presently be detected even on a global scale."[12] That makes the conclusion of the 2001 IPCC report even less surprising: if rainfall poses a challenge for identifying a greenhouse signal, then we should expect it to be very difficult to find a greenhouse signal in stream flows of rivers and consequent flood damage.

In many locations, however, trying to tie rainfall and stream flow to losses due to floods is somewhat beside the point. That is to say, flood losses have not increased at the same rate that development has. This suggests that flood policies are having a positive impact on economic losses, or possibly even that climate has become more benign, or some combination of the two. Either way, research on flood losses has found that development in exposed regions plays the major role in explaining increasing damage.

We found something a bit more titillating, too: politics has a strong impact on the number of events officially declared disasters. We examined a database of U.S. federal disaster declarations, which occur when the governor of a state asks the president for monetary assistance to help respond to a disaster that overwhelms local and state capabilities. The trend in disaster declarations has been sharply upward since the 1980s, leading some to see a signal of human-caused climate change in the trend. Our research suggests a simpler explanation, however. For one thing, U.S. government policy has become increasingly generous. Federal money just flows more freely after disasters these days than it did in years past. Decisions about declaring disasters and allocating government funds of course have a strong political dimension. Consider that presidents declare 50 percent more flood disasters in years in which they are running for reelection than in other years. There is obviously a signal there. It's just not a climate signal.

Research on floods does not support the hypothesis that greenhouse gas emissions have led to a discernible increase in flood losses. Hans Jochim Schellenhuber, a prominent climate scientist and scientific adviser to German chancellor Angela Merkel, coauthored a 2004 paper that summarized the 2001 IPCC assessment and subsequent scientific discussions of floods. It concluded: "There has been no conclusive and general proof as to how climate change affects flood behavior, in the light of data observed so far. . . . It is difficult to disentangle the climatic component from the strong natural variability and direct, man-made, environmental changes."[13]

While flood events around the world are often linked by the media and activists to greenhouse gas emissions, the scientific research on floods and their damage shows no basis for making such a connection. It is today not possible to identify the influence of accumulating carbon dioxide (and other greenhouse gases) in the atmosphere in the global disaster record of hurricanes and floods, much less in any region around the world; the group we gathered at Hohenkammer in 2006 thought an unequivocal link was unlikely to be found anytime soon, and that consensus has been reinforced by the research published in the four years since.

Other sorts of extreme events—such as heat waves, cold spells, droughts, and forest fires—have similarly complex stories. In all cases, the societal part of the equation that leads to losses is the dominant factor, not climate change. Yet for some phenomena there are indications that a signal of climate change, and possibly related to greenhouse gas emissions, can be seen in the record of impacts. For instance, some recent research is suggestive that regional warming in the western United States can be associated with increasing forest fires, even in the context of complex patterns of forest management over the past century.[14] If that pattern of warming can be directly attributed to greenhouse gas emissions, then it would be possible to attribute human-caused climate change with a demonstrated impact.

Other regional impacts illustrate why such attribution is difficult. A recent regional drought that affected the U.S. Southeast, particularly Atlanta, was used by some in the climate debate to advocate for changes to emissions policies; however, subsequent research suggests that the drought was fairly typical but that its impacts were much more significant due to societal development and the corresponding increased demand for water.[15] Similarly, research that I am now involved in with colleagues at Macquarie University in Sydney indicates that the horrific losses of the 2009 bushfires in Victoria, Australia, were not outside the bounds of what should be expected, based on historical events adjusted for societal change. As the climate changes, it is likely that scientists will detect more and more instances where those changes have had discernible effects on the impacts that people care about. At the same time, the ability of scientists to attribute those impacts conclusively and directly to greenhouse gas emissions will likely remain limited.

Teasing out a signal of greenhouse gas effects on climate impacts will remain difficult because there are not good data over time on the impacts of these phenomena, and also because of the complexity of the various factors that shape loss trends. Of course, if the point is to see if greenhouse gases have affected extreme weather events, looking specifically at climate data will always make more sense than trying to see such effects in the impact record. A recent study of potential changes in hurricane behavior suggests that greenhouse gases will lead to fewer but possibly more intense hurricanes over the coming century. The study also concluded that even if such changes occur, they would not even be detectable until the latter half of the century, due to the considerable background variability in the climate system.[16] If it takes that long to detect a change in the physical climate system, it is a certainty that it would take considerably longer to detect a change in the corresponding impacts to society or ecosystems.

It is important to underscore the fact that just because a greenhouse gas signal has not yet been detected for a given phenomenon, such as hurricanes or floods, does not mean that one never will be. Our inability to detect changes does not mean that changes are not taking place or won't in the future. However, what does seem to hold across all these situations is that the signal of climate (much less a signal from human-caused climate change) can be very difficult to see in the context of everything else that leads to impacts on human systems from extreme events. From a policy perspective this means that the most important policy levers will necessarily be adaptive, for the simple reason that such policies can influence losses immediately. Nonetheless, disasters are routinely exploited in climate advocacy in support of emissions reductions and, as we shall see, even in leading scientific assessments.

Losing Touch with Empirical Science: Guessing Games and Health Effects

Much has been made of the possibility of links between increased greenhouse gas emissions and disease, although little of it is concrete. In 2002 the World Health Organization concluded that it could not identify the influence of greenhouse gas emissions on health and disease based on existing data: "Climate exhibits natural variability, and its effects on health are mediated by many other determinants. There are currently insufficient high-quality, long-term data on climate-sensitive diseases to provide a direct measurement of the health impact of anthropogenic climate change, particularly in the most vulnerable populations."[17]

A year later, however, another WHO study argued that even though the data did not allow scientific conclusions about the influence of greenhouse gas emissions on human health, speculative guesses were necessary. Waiting to gather better data "may accord with the canons of empirical science," the WHO report admitted, but "it would not provide the timely information needed to inform current policy decisions on GHG [greenhouse gas] emission abatement, so as to offset possible health consequences in the future."[18] That is, they wouldn't be able to contribute to political debates about emissions reductions if they didn't make any claims about climate change and disease.

The speculative guesses of WHO formed the basis of estimates released in a 2009 report issued by the Global Humanitarian Forum, a nongovernmental organization run by former UN secretary-general Kofi Annan. The GHF concluded that greenhouse gas–driven climate change was presently responsible for 154,000 deaths per year due to malnutrition, 94,000 deaths per year due to diarrhea, and 54,000 deaths per year due to malaria, which when added to deaths from weather-related disasters (which have declined dramatically over the past century) gives a total of 315,000 people who allegedly die each year due to human-caused climate change. A close look at the health-related numbers shows that they are exactly two times the values presented in the 2002 WHO report, which said that the estimates do not "accord with the canons of empirical science."[19] In other words, the numbers appear to be just a guess on top of the earlier speculation. "Analyses" such as these are what give some areas of climate science a bad name and suggest an unhealthy politicization of research to support favored causes.

But let's assume for a moment that the speculation were true, and put it in context. The WHO estimates that in 2004, 2.16 million people died from diarrhea, 888,000 from malaria, and 485,000 from nutritional deficiencies.[20] If one generously takes as true the estimate of about 300,000 deaths per year directly due to the consequences of greenhouse gas emissions on diarrhea, malaria, and malnutrition, then efforts to mitigate climate change, if 100 percent effective and with the immediate effect of reducing deaths, could at best address less than 10 percent of the total global deaths due to these causes. By contrast, efforts to address root causes of diarrhea, malaria, and malnutrition offer the prospect of saving millions of lives, including those that may be affected by human-caused climate change. Further, one need not have any certainty as to the exact role of greenhouse gas emissions in exacerbating human health effects to understand that adaptive responses offer the greatest hope for progress with respect to reducing mortality. The treatment of the health impacts of climate change is troubling. In the context of disasters, unfortunately, it gets even worse.

Untangling a Decade of Misrepresentation of Disasters and Climate Change

In 2001 the IPCC Third Assessment Working Group II report cautiously claimed that the upward trend in the costs of disasters had a climate component. The IPCC supported this assertion by referencing a short report by Munich Reinsurance published in 2000 that surveyed natural disasters in 1999.[21] The IPCC reproduced a version of Figure 7.1 in the report, explaining that Munich Reinsurance "estimates that economic losses from large natural disasters increased twofold between the 1970s and 1990s, after correcting for inflation, insurance penetration and pricing effects, and increases in the material standard of living."[22] The IPCC only cautiously speculated on issues of why this increase had occurred, naming both socioeconomic factors and climatic ones—but never climate changes due to greenhouse gas emissions.

A quick look at the data explains why the IPCC's caution was appropriate. The Munich Reinsurance report relied on by the IPCC concluded that global disasters resulted in $636 billion in losses in the 1990s compared with $315 billion in the 1970s, after adjusting for changes in population and wealth.[23] Even though the Munich Reinsurance data included earthquake disasters and other non-climate-related losses, the IPCC suggested that "climatic factors"—broadly speaking, and not simply because of change due to greenhouse gases—must therefore play some part in the increase.

Even such a cautious claim needs to be carefully examined. The large decadal variability in disaster losses makes it quite dodgy to assert a trend by comparing two different ten-year periods over a total period of only thirty years. This can be illustrated with an example from our database of hurricane losses. If we use the hurricane-loss data, accounting for trends in population, wealth, and inflation, from 2005 values (as shown above in Figure 7.6) and then compare decades, as was done in the Munich Reinsurance report, we see some interesting things. First, the ratio of losses in the 1990s to those of the 1970s is 2 to 1, which is almost identical to the Munich Reinsurance Group's analysis for global losses. The closeness of ratios should not be surprising, because U.S. hurricane losses are the largest contributor to the Munich Reinsurance global data set. But if we look at other decadal comparisons of the twentieth century, the picture looks quite different: the 1990s had only 80 percent of the normalized losses of the 1940s and less than 50 percent of the normalized losses of the 1920s. Going all the way back to the start of the data set in the decade of the 1900s, losses were just about the same as in the 1990s. The choice of starting point makes a huge difference, and over the longer time period in the hurricane loss data, as well as that for other phenomena around the world, there is simply no evidence of an increase in disaster losses beyond that which is explained by development and economic growth. The bottom line is that the 2000 Munich Reinsurance Group's analysis tells us absolutely nothing about the attribution of the causes of increasing disaster losses. And in 2001 the IPCC reported this quite appropriately.

However, the IPCC's caution was not followed by everyone who subsequently relied on its analysis. In 2005 *Science* published a commentary by Evan Mills, of the U.S. Department of Energy's Lawrence Livermore National Laboratory; the commentary cited the 2001 IPCC report. In his essay, Mills asserted boldly and incorrectly that the IPCC assessment showed that "climate change has played a role in the rising costs of natural disasters" and attributed at least part of the increase to "anthropogenic climate change." In addition to incorrectly citing the IPCC as providing support for his claim that "anthropogenic climate change" was a factor behind increasing disaster losses, Mills also cited separately the Munich Reinsurance 2000 report that was the exact same reference cited by the IPCC, giving the impression that there were multiple independent sources of support for his claim. (Even worse, he cites a third report that also relies on the Munich Reinsurance report.)[24] Mills's missteps would not be important except for the fact that, as we will shortly see, due to its prominence in the prestigious journal *Science*, his 2005 commentary is oft cited in support of claims that increasing disaster losses can be attributed to greenhouse gas emissions.

Understanding the systematic misrepresentation of the science of climate change and disaster losses is a bit like trying to decipher the children's game of "Chinese whispers" or "telephone," in which a child whispers something to his or her neighbor and so on around a circle; what is reported to the last child often bears no resemblance to that started by the first. For instance, in 2006 the widely influential Stern Review Report on the Economics of Climate Change, produced by the government of the United Kingdom, relied on Mills's 2005 paper and a second paper to claim that greenhouse gas emissions were causing disaster losses to increase by 2 percent per year. That second paper Stern cited was very familiar to me; it was one of the white papers prepared by a participant in our 2006 Hohenkammer disaster-loss workshop. That paper was written by Robert Muir-Wood (and colleagues) of Risk Management Solutions, a catastrophe modeling company that provides estimates of disaster risks and losses to insurers and reinsurers.[25] Muir-Wood's paper found that if you start your analysis of disaster losses in the 1970s, when costly disasters were at a century-long low, you will find an increase over recent decades, especially (as Muir-Wood noted) if the last year in your data set is 2005, when Hurricane Katrina struck. Muir-Wood and colleagues found no trend in adjusted losses if you start the analysis in the 1950s. A related, subsequent paper by Muir-Wood and colleagues concluded, "We find insufficient evidence to claim a statistical relationship between global temperature increase and normalized catastrophe losses."[26]

Remarkably, not only did the Stern Review ignore the growing peer-reviewed literature on disasters and climate change, but within the Muir-Wood analysis the review selectively used the shorter time frame to generate an estimate of escalating damages due to greenhouse gas emissions.[27] In early 2010 Muir-Wood was scathing in his criticism of the Stern Review for misusing his research, saying it went "far beyond what was an acceptable extrapolation of the evidence."[28]

The full report from our Hohenkammer workshop as well as the workshop's consensus statements (signed onto by Muir-Wood) directly contradicted the Stern Review's analysis; however, neither were cited in the review. Motivated by this obvious misrepresentation of the science of disasters and climate change, I examined the Stern Review in depth and found several additional and significant errors in its treatment of disasters and climate change, including an apparent uncaught typo in the economic effect of hurricane damages that inflated them by an order of magnitude.[29] The effects of various errors and mistakes in the review's total estimate of the economic damage from human-caused climate change add up to as much as 40 percent of the review's total estimated losses from all of the effects of greenhouse gas emissions (a number calculated using a separate methodology), not an insignificant amount.[30]

The following year, in 2007, the IPCC released its Fourth Assessment Report, and it also relied on the Muir-Wood study from our Hohenkammer workshop as the "one study" to highlight in its summary review of disasters and climate change. Ignoring the longer period looked at by Muir-Wood as well as the analyses we did of the entire twentieth century, the IPCC selectively concluded that "once losses are normalized for exposure, there still remains an underlying rising trend."[31] Even worse, the IPCC included in its supplementary material a graph that plotted temperature alongside disaster losses, smoothing the data and scaling the axes in such a way as to suggest a relationship, despite the fact that none had been shown in the peer-reviewed literature. The figure is cited indicating that the Muir-Wood study from our workshop is the basis for the graph, although no such analysis appears in that paper.[32] In early 2010 Muir-Wood revealed that the IPCC knowingly miscited the graph, which he created informally and says was a mistake to include in the IPCC report, in order to circumvent a deadline for inclusion of materials in the report.[33] Ironically, the paper that the IPCC actually wanted to cite (but could not because of the deadline) found "insufficient evidence to claim a statistical relationship between global temperature increase and normalized catastrophe losses."[34]

Furthermore, the IPCC somehow neglected to mention the many other peer-reviewed studies examining a wide range of places and time periods that found no signal of anthropogenic climate change after adjusting for societal factors, and while it cited our workshop report, it failed to report its conclusions about the present impossibility of attributing disaster losses to greenhouse gases. Either the IPCC was very sloppy, or it went to great lengths to suggest in a misleading manner a connection between rising temperatures and increasing disaster losses—or both.

It turns out that several reviewers of the IPCC report had in fact raised questions about its treatment of the issue of disaster losses. One reviewer questioned the IPCC's suggestion that our normalization work had been superseded by events, and asked the IPCC directly, "What does Pielke think about this?" The IPCC responded on my behalf, explaining, "I believe Pielke agrees that adding 2004 and 2005 has the potential to change his earlier conclusions—at least about the absence of a trend in US Cat[astrophe] losses."[35] The problem with the IPCC response to the reviewer was that it was a complete fabrication. Just two months before I had published a paper with a version of Figure 7.6 (see page 171) showing clearly that the events of 2004 and 2005 did not change the overall picture at all.[36] The IPCC included misleading information in its report and then fabricated a response to a reviewer, who identified the misleading information, to justify keeping that material in the report.

By that time I had become accustomed to the misrepresentation of disaster losses and climate change. So when the U.S. Climate Change Science Program released a report on climate extremes under the Bush administration in 2008 and then a related synthesis report under the Obama administration in 2009, I was not at all surprised to see what had become a familiar misdirection in the reports. The 2008 CCSP report on North American extremes cited several of my papers (and one on tornadoes by others using a variant of our normalization methodology) in support of claims that these papers did not make. Specifically, the report claimed that increasing disaster losses were due to both societal and climatic factors, as follows: "Numerous studies indicate that both the climate and the socioeconomic vulnerability to weather and climate extremes are changing" and cited three sources in support of this sentence.[37] None of the three studies cited supports the claim being made of an increase in losses due to climate change. In fact, they all say the opposite:

Brooks and Doswell: "We find nothing to suggest that damage from individual tornadoes has increased through time, except as a result of the increasing cost of goods and accumulation of wealth of the United States."

Pielke et al.: "The lack of trend in twentieth century normalized hurricane losses is consistent with what one would expect to find given the lack of trends in hurricane frequency or intensity at landfall."

Downton, Miller, and Pielke, which does not analyze the role of climate: "[U.S. flood] damage per unit wealth has declined slightly."

Further, the Bush administration's CCSP report relied on Evan Mills's 2005 article in *Science* as a source for claims of attribution. Remarkably, Mills was an author of the CCSP report that evaluated his own work. Given what the body of the CCSP report said about trends in extreme events in North America (see "U.S. Extreme Events," page 184), its ultimate conclusion that climate change (regardless of cause) is behind trends in increasing losses makes for a big self-contradictory surprise.

The CCSP subsequently issued a synthesis report under the Obama administration. It engaged in the same misrepresentation of the peer-reviewed literature, concluding, "While economic and demographic factors have no doubt contributed to observed increases in losses, these

U.S. Extreme Events: Excerpts from a 2008 U.S. Government Report

1. Over the long term U.S. hurricane landfalls have been declining.

"The final example is a time series of U.S. landfalling hurricanes for 1851–2006. . . . A linear trend was fitted to the full series and also for the following subseries: 1861–2006, 1871–2006, and so on up to 1921–2006. As in preceding examples, the model fitted was ARMA (p,q) with linear trend, with p and q identified by AIC.

"For 1871–2006, the optimal model was AR(4), for which the slope was –.00229, standard error .00089, significant at $p = .01$. For 1881–2006, the optimal model was AR(4), for which the slope was –.00212, standard error .00100, significant at $p = .03$. For all other cases, the estimated trend was negative, but not statistically significant."

2. Nationwide there have been no long-term increases in drought.

"Averaged over the continental U.S. and southern Canada the most severe droughts occurred in the 1930s and there is no indication of an overall trend in the observational record."

3. Despite increases in some measures of precipitation there have not been corresponding increases in peak stream flows (high flows above 90th percentile).

"Lins and Slack (1999, 2005) reported no significant changes in high flow above the 90th percentile. On the other hand, Groisman et al. (2001) showed that for the same gauges, period, and territory, there were statistically significant regional average increases in the uppermost fractions of total streamflow. However, these trends became statistically insignificant after Groisman et al. (2004) updated the analysis to include the years 2000 through 2003, all of which happened to be dry years over most of the eastern United States."

4. There have been no observed changes in the occurrence of tornadoes or thunderstorms.

"There is no evidence for a change in the severity of tornadoes and severe thunderstorms, and the large changes in the overall number of reports make it impossible to detect if meteorological changes have occurred."

5. There have been no long-term increases in strong East Coast winter storms (ECWS), called nor'easters.

"They found a general tendency toward weaker systems over the past few decades, based on a marginally significant (at the $p = 0.1$ level) increase in av-

(continues)

erage storm minimum pressure (not shown). However, their analysis found no statistically significant trends in ECWS frequency for all nor'easters identified in their analysis, specifically for those storms that occurred over the northern portion of the domain (>35°N), or those that traversed full coast (Figure 2.22b, c) during the 46-year period of record used in this study."

6. There are no long-term trends in either heat waves or cold spells, though there are trends within shorter time periods in the overall record.

"Analysis of multi-day very extreme heat and cold episodes in the United States were updated from Kunkel et al. (1999a) for the period 1895–2005. The most notable feature of the pattern of the annual number of extreme heat waves (Figure 2.3a) through time is the high frequency in the 1930s compared to the rest of the years in the 1895–2005 period. This was followed by a decrease to a minimum in the 1960s and 1970s and then an increasing trend since then. There is no trend over the entire period, but a highly statistically significant upward trend since 1960. . . . Cold waves show a decline in the first half of the 20th century, then a large spike of events during the mid-1980s, then a decline. The last 10 years have seen a lower number of severe cold waves in the United States than in any other 10-year period since record-keeping began in 1895."

Source: http://downloads.climatescience.gov/sap/sap3-3/sap3-3-final-all.pdf.

factors do not fully explain the upward trend in costs or numbers of events." It used our 2008 hurricane paper to support the claims made in the first half of the sentence. The second half was supported by two references. One was Mills's 2005 article, which, as we've seen, traces its (unsupportable) claims through the 2001 IPCC to the 2000 Munich Reinsurance report comparing losses from the 1970s to the 1990s. The other was the 2007 IPCC report, which, as we have seen, relies exclusively on the Robert Muir-Wood white paper contributed to our Hohenkammer workshop. Muir-Wood's work actually concluded that no climate signal could be found in the adjusted losses over the period of its analysis.

The vignettes presented above document a pattern of unsupportable and just plain incorrect representations of the science of disasters and climate change in the Stern Review Report, the reports of the

IPCC, and the U.S. government climate science reports. The pattern has three common characteristics:

1. Reliance on a small number of non-peer-reviewed, scientifically unsupportable studies rather than the relevant peer-reviewed literature, which tells a far different story.
2. Reliance on and featuring non-peer-reviewed work conducted by the authors of the assessment reports. Questions raised in the review process of these reports were ignored or were responded to with incorrect and misleading information.
3. Repeated reliance on a small number of secondary or tertiary sources, repeatedly cited such that intellectual provenance is lost.

Finally, in late 2009 and early 2010 a range of issues came to light involving sloppiness and errors in the reports of the Intergovernmental Panel on Climate Change. While some reflexively dismissed the allegations, environmental ministers from around the world saw enough merit in the claims to request an independent review of the policies and procedures of the IPCC. This was not only to improve its quality control but also to restore some of the credibility that the institution had lost due to the errors and the ham-handed IPCC response. The IPCC strongly denied any problems whatsoever in its handling of the issue of disaster losses and dismissed claims of breakdowns as baseless in a strongly worded press release.

Making Sense of Misrepresentation

Some participants in the debate over climate change, especially those opposed to action, are quick to see base political motives at play in the mistakes found in the IPCC (and the Stern Review and CCSP) and the defensive circle-the-wagons approach. They claim that a desire to advance the cause for action has led to willful misrepresentations of aspects of the science, with some going so far as to suggest that all of climate science is a hoax or fraud.

Such views are nonsense. As I detailed in Chapter 1, the science supporting claims of a human influence on the climate system is robust, and if anything, the diversity of human influences on the climate has been underestimated by the IPCC, thanks to its fixation on carbon dioxide. Although there have been problems in the IPCC, understanding why they have occurred requires considerably more sophistication than claims of politically inspired fraud or hoax.

Together, the more conceptual arguments of Chapter 6 and the more analytical arguments found earlier in this chapter help to explain the incentives that underlie the making of unsupportable claims about the connection of greenhouse gas emissions and disasters, as well as other impacts, such as on human health. The narrow definition of climate change used by the Climate Convention makes the identification of a "dangerous interference" in the climate system due to greenhouse gas emissions a requisite criterion for action. If there was no such interference, then action would be unnecessary. For those wanting to argue a case for action, there are strong incentives to attribute impacts of concern to society (and disasters are of utmost concern) to greenhouse gas emissions. Conversely, statements that such impacts cannot presently be seen, or that they may be very difficult to see for many decades into the future, are difficult to square with the idea of dangerous interference. Thus, the policy framework itself creates incentives to view the science in a particular way. The IPCC, Stern Review, and CCSP are all institutions that have been put into place to support the case for action on climate change.

There is, of course, no excuse for the IPCC's misrepresentation of my views in the review process, or for the misreferencing of a paper to avoid a publication deadline, or similarly for the Stern Review and CCSP to mischaracterize published studies. Nevertheless, even if we can't excuse it, we can explain it. Robert Watson, former director of the IPCC, expressed concern that each error and instance of sloppiness produced by the IPCC's working group on impacts has "overstated the implications of climate change."[38] Indeed, the broader pattern of misrepresentation of the science of climate impacts—what might be generally characterized by using the terminology of WHO as departing from the "canons of empirical science"—reflects selectivity, bias, and a systematic shading of the evidence to fit into a policy framework put in place under the Climate Convention. Judy Curry, a climate scientist at Georgia Tech, explains how the Climate Convention influences presentations of the science both in the IPCC but more broadly as well: "The UNFCCC has a particular policy agenda—Kyoto, Copenhagen, cap-and-trade, and all that—so the questions that they pose at the IPCC have been framed in terms of the UNFCCC agenda. That's caused a narrowing of the kind of things the IPCC focuses on. It's not a policy-free assessment of the science. That actually torques the science in certain directions, because a lot of people are doing research specifically targeted at issues of relevance to the IPCC. Scientists want to see their papers quoted in the IPCC report."[39]

While such tactics are to be expected from advocates for action on all sides of the debate, it is more than a little troubling to see authoritative scientific institutions playing these kinds of games. For those who believe that accurate information wins out in the long run, it should be obvious that misrepresentations of the science of climate impacts cannot be sustained. Lost credibility is inevitable when the facts do come to light.

As Chapter 6 argued, which is worth repeating, the policy framework of the Climate Convention is almost entirely about mitigation, despite its recognition of the need for adaptation. Adaptation is defined to be a cost of failed mitigation, and thus is to be avoided. With complete success in mitigation under the logic of the Climate Convention, adaptation would be unnecessary, because there would be no dangerous interference in the climate system. In this context, the subtext of scientific assessments supposedly about adaptation is often really about mitigation, and specifically about providing support for claims of a "dangerous interference."

Adaptation from a Broader Perspective

It should be clear from the analysis in this chapter that adaptation has important dimensions that are independent of energy policy. To call for a greater focus on adaptation is not to support business-as-usual energy policies, or to be interpreted in any way as counter to climate mitigation. Adaptation and mitigation are not trade-offs but complements that address different issues on very different timescales of costs and benefits. If a policy goal is to reduce the future impacts of climate on society, then energy policies are insufficient, and indeed largely irrelevant, to achieving that goal. There are other sensible reasons for efforts to accelerate decarbonization; protecting us from disasters is not one of them, and arguments and advocacy to the contrary are not in concert with research in this area.

Governments and businesses are already heavily invested in climate policy and thus should focus resources on decisions likely to be effective with respect to policy goals. In the context of extreme events, such decisions might focus increasingly on land use, insurance, engineering, warnings and forecasts, risk assessments, and so on. These policies can make a large difference in mitigating the future impacts of climate on society.

If we change the context, the politics fall away, and we are better able to see the forest for the trees with respect to adaptation. Consider the massive earthquakes that struck Haiti and Chile in early 2010. Despite the fact that both earthquakes struck in highly populated regions, and the fact that the Chilean earthquake was many times more powerful than the one that shook Haiti, human losses in Chile were far less than the horrific loss of life experienced in Haiti. A U.S. Geological Survey expert explained: "The standard mantra is earthquakes don't kill people, buildings do."[40] Just as with earthquakes, with respect to weather extremes, where, how, and what we build are the driving factors underlying trends in losses. The intense politics of the climate debate just makes this reality a bit more difficult to see, as compared to the far less politicized issue of earthquakes.

Looking to the future, improved policies in the face of climate extremes share much in common with policies in anticipation of earthquakes. The living conditions of those exposed to extremes are by far the most important factor in the losses that will be experienced in the future. Given that many communities around the world, especially in poor countries, lack a basic resilience in the face of climate extremes, there is much work to be done to improve adaptive capacities. Such policies make sense independent of human-caused climate change, but they will also make these communities more robust in the face of human-caused climate change. Adaptation involves the fuller realization of objectives for human dignity, including security, well-being, and prosperity. Rather than seeing adaptation as simply the cost of failed climate policies, adaptation deserves to be at the center of a positive, human-focused agenda that is independent of carbon policy. Even Al Gore has come around to this view: "I used to think adaptation subtracted from our efforts on prevention. But I've changed my mind. Poor countries are vulnerable and need our help."[41] He is right.

CHAPTER 8

The Politicization of Climate Science

THE SYSTEMATIC misrepresentation of the science of climate impacts documented in the previous chapter only scratches the surface of the politicization of climate science. Such politicization manifests itself in the actions of climate scientists, in the presentation of climate science results to policy makers, and in the structuring of scientific research and assessments. Climate science is today a fully politicized enterprise, desperately in need of reform if integrity is to be restored and sustained.

In November 2009 someone stole or leaked more than a thousand e-mails and various other documents from the Climatic Research Unit of the University of East Anglia in the United Kingdom. Even before the authenticity of the e-mails was verified, the materials were widely disseminated across the Internet. The private communications—dating back to the 1990s—among leading climate scientists were extremely troubling to many commentators. For instance, George Monbiot, an environmental campaigner and columnist for *The Guardian* newspaper in the UK, wrote upon first seeing them, "It's no use pretending that this isn't a major blow. . . . I'm dismayed and deeply shaken by them."[1]

What was it that was so troubling in the e-mails? For some it was the presence of eyebrow-raising, but ultimately somewhat ambiguous, language, such as discussing a "trick" to "hide the decline" and the "travesty" that "we can't account for the warming at the moment." The scientists involved and their defenders were quick to respond that the e-mails just showed scientists hard at work: of course the e-mails were sometimes harsh and blunt, they countered, but what could anyone expect from scientists being hounded by climate skeptics and deniers? Others saw in the e-mails the crumbling of the entire edifice of climate science, designating the event "Climategate."

As with most everything in the climate debate, reality is more nuanced and complex than such simple interpretations. Despite claims to the contrary by some vocal commentators, the issues revealed by the e-mails were not about scientific misconduct or willful fraud.[2] To the contrary, the issues were far more about norms of behavior rather than malfeasance, a distinction that was aptly summarized by Clive Crook, a columnist for the *Financial Times*, from a perch far removed from debates over climate science:

> Any fair-minded person would regard those [East Anglia e-mail] exchanges as raising questions. On the face of it, these are not the standards one expects of science. Nor is this just any science. The work of these researchers is being used to press the case for economic policies with colossal adjustment costs. Plainly, the highest standards of intellectual honesty and openness are called for. The emails do not attest to such standards. Yet how did the establishment respond? It said that this is how science is done in the real world. Initially, the head of the Intergovernmental Panel on Climate Change defended the scientists and played down the significance of their correspondence.[3]

What standards of intellectual openness and honesty had the authors of the e-mails compromised? For one thing, the scientists—who saw themselves as much as activists as researchers—expressed a desire to game the system of peer review in scientific publishing such that their opponents were denied a chance to publish their work in scientific journals or have their work cited in the reports of the IPCC while the scientists' allies had a much easier time. Publication in peer-reviewed journals matters for people's careers, of course, but in the case of climate science, findings should (in principle) be published in peer-reviewed publications as a prerequisite to being highlighted in the IPCC assessment of climate science, arguably the most authoritative and influential summary of the state of research.

Peer review in scientific publishing is a process in which experts are asked to judge the appropriateness of a paper for publication in a scientific journal. It is often cursory and focused on the merits of an argument, rather than a detailed replication of the analysis or decomposition of the methods. Peer review does not mean that a result is right or will stand the test of time; rather, it has met some minimal standards of acceptability for publication. The scientific community is replete with vignettes about papers that were rejected for publication in one venue only to be published elsewhere and later turned out to be seminal. Similarly, every so often even *Science* and *Nature*, the most prestigious science journals, find themselves in trouble for having published a paper that is badly wrong or even fraudulent. But despite these shortcomings in the process, peer review is widely viewed much as Winston Churchill viewed democracy: the worst possible system except for all the others.

Peer review in science works because over the long-term good ideas and solid arguments win out. This process happens organically through a decentralized process. Peer review takes place through many different and independent scientific journals, with editing and reviewing conducted by many independent scholars from a range of disciplinary and experiential backgrounds, and with their own idiosyncratic biases and views. No one group or perspective owns the peer-review process, and the diversity of the scientific enterprise is part of its core strength. Truth—meaning a convergence to agreement on scientific questions—thus is a product of the peer-review process over time.

Of course, the path to truth can be convoluted and indirect. And scientists don't always reach agreement, particularly on timescales of relevance to decision makers. Consequently, complex issues relevant to important decisions are often characterized by uncertainties, contested certainties, and fundamental areas of ignorance; combined, those factors can make the distribution of scientific views not readily apparent, even to the informed observer. In such situations a formal assessment can provide a useful perspective on the degree of consensus or disagreement among relevant experts on various claims. Such assessments are nothing more than a snapshot in time, as science is continuously evolving. When done well, an assessment will reflect the full range of views held by relevant experts, including minority views, as well as the connections of scientific understandings to alternative possible courses of action. When done poorly—either because minority views are kept out of the scientific press to begin with or because they're excluded from the process of writing an assessment—the documents can be misleading or damaging.

The "Climategate" e-mails show a consistent desire among the activist scientists to redefine processes of peer review in accordance with their own views of climate science, such that papers that supported their views would appear in the literature and those that did not would be rejected. If successful, such a redefinition would turn the entire notion of peer review on its head. The e-mails indicate concerted efforts to reshape the peer-review process by managing and coordinating reviews of individual papers, by putting pressure on journal editors and editorial boards, by seeking to stack editorial boards with like-minded colleagues, by arranging boycotts of journals, and through other unprofessional actions. Why? Because of the short-term politics of climate change.

The scientists exposed by the East Anglia e-mails apparently decided that the peer-review process would work better in service of their political agenda (which was focused on defeating the "skeptics" in public debate) if they used "truth" to determine whose views would be allowed to be published in the literature and reflected in assessments. In this case "truth" simply means the views deemed acceptable among these activist scientists and their close clique of colleagues. NASA climate scientist Gavin Schmidt, who was among those whose e-mails were revealed, defended this very backward view of peer review in terms of the implications for climate politics: "In any other field [a bad paper] would just be ignored. The problem is the climate field has become extremely politicized, and every time some nonsense paper gets into a proper journal, it gets blown out of all proportion." Schmidt explained that the papers that he and his colleagues judged to be low quality were nonetheless being read and discussed by skeptics of the science of climate change.[4] So Schmidt appears to be suggesting that in order to limit the ability of their political opponents to cherry-pick and blow out of proportion studies that the activist scientists did not agree with, they saw a convenient shortcut: simply reshape the peer-review system such that papers that they disagree with never get published in the first place, or at least are never mentioned in scientific assessments. If those papers were unpublished or unmentioned, then they would have far less impact in the public debate on climate change.

The problem with this strategy, of course, is that many climate scientists (and presumably others inside and outside of the scientific establishment) are unwilling to cede ownership of "truth" to a small clique of scientists. Peer review exists in the first place because there are no shortcuts to the truth, and any such shortcut will inevitably fail. Consider that the efforts revealed in the East Anglia e-mails to manage the peer-reviewed literature went well beyond efforts to prevent so-called skeptical papers from being published, but included a focus on papers that fully accepted a human influence on climate but offered views that differed in some degree from those preferred by the activist scientists.[5] The e-mails reveal activist scientists busy extolling the virtues of peer review to journalists and the public, while at the same time they were busy behind the scenes plotting to corrupt the peer-review process in a way that favored their views on the science and politics of climate change.

The clique of activist scientists sees absolutely nothing wrong in what it was doing—after all, it was just about enlisting the "truth" in support of the greater good. And the issue is made even more complex because those who share the political agenda of the activist scientists are ready to join their coup against peer review, whereas those opposed to their political agenda are happy to try to exploit the ethical lapses of the scientists and to criticize the activist scientists for politicizing climate science, as somehow being proof of the "skeptical" position on climate change. So much of the discussion gets wrapped up in these maneuvers for political gain, rather than the issue of the integrity of climate science. Rather than bringing science more fully to bear on the complex politics of climate change, the actions of these activist climate scientists did far more to bring the messy world of climate politics into the enterprise of climate science. Politics has also found its way into climate science in much more subtle ways, which, although less dramatic, are no less important for understanding how climate science has become worryingly politicized.

The Fear Factor

One of the most common tactics used in the climate debate in an effort to motivate action is use of fear, especially when such strategies are sanctioned by climate scientists. The thinking behind the fear factor is that if people are alarmed about the prospect of human-caused climate change, they will then be motivated to support action in response, with, the argument holds, more fear leading to more alarm, leading to greater support for action, and ultimately the desired action itself. Reality, however, does not work like this.

The use of fear is consistent with the notion that we lack political will (discussed in Chapter 2) and the idea that political will is created through acceptance of a particular view of the science of climate change. The use of fear is also consistent with efforts to demonstrate "dangerous interference" as encouraged under the Climate Convention (and discussed in Chapter 6). The underlying logic here is that once someone accepts the science, they then inevitably and inexorably come to accept a specific set of policy solutions.

Consider the views of Professor John Schellenhuber, of the Potsdam Institute for Climate Impact Research in Germany and an adviser on climate issues to German chancellor Angela Merkel: "The U.S. in a sense is climate illiterate. It is a deeper problem in the U.S., if you look at global polls about what the public knows about climate change. Even in Brazil and China, you have more people who know the problem, who think that deep cuts in emissions are needed." The logic implied by Professor Schellenhuber's reasoning is obvious: not enough of the public actually understand the science of climate change, and thus, because of the public's lack of understanding, the U.S. government has not taken action on climate change.[6] Consequently, the implied remedy is both to better educate the public about climate change and to defeat those who express skepticism about the science in public debate—that is, those who are opponents to the "proper" understanding. We've already seen in Chapter 2 that logic like Schellenhuber's is unsound: opinion polls over more than a decade show strong public support for action in the United States and stronger support elsewhere. Regardless, because advocates like Schellenhuber think that one's "correct" or "incorrect" politics flows from the correctness of one's views of science, climate science has become fully politicized, as, for many participants, the debate has been framed such that debates over science are equivalent to debates over politics. Hence, the issue ultimately devolves to an absurd end point, such as when, in a November 2009 speech, Australia's prime minister, Kevin Rudd, made no distinctions between those whom he denigrated as "climate deniers" who reject certain aspects of the science of climate change and those who accept these aspects of science but disagree with his favored political course of action.[7]

Mike Hulme explains the general approach to perceptions of an "unreasonable" public: "One reaction to this 'unreasonableness' is to get scientists to speak louder, more often, or more dramatically about climate change. Another reaction from government bodies and interest groups is to use ever-more-emotional campaigning."[8] For instance, the Copenhagen climate conference opened with a short film titled *Please Help Save the World* about a child suffering the consequences of a global apocalypse that would seem to be more appropriate in campy big-screen science fiction than a sober global negotiation.

This framing of the political challenge has led to the debate's being characterized by supporters of action on climate change as a battle between those who are guided by science, on the one side, and the ignorant, antiscience deniers, on the other. The central logic here is an assumption that once people come to understand what the science really says, they will then support the associated course of action. That this line of thinking reflects what has been called the naturalistic fallacy—the false assumption that you can get an "ought" from an "is"—has not stopped the debate from being framed and carried out in this manner.

Typically, the battle over climate-change science focused on convincing (or, rather, defeating) those skeptical has meant advocacy focused on increasing alarm. As one Australian academic put it at a conference at Oxford University in the fall of 2009: "The situation is so serious that, although people are afraid, they are not fearful enough given the science. Personally I cannot see any alternative to ramping up the fear factor."[9] Similarly, when asked how to motivate action on climate change, Nobel Prize–winning economist Thomas Schelling replied, "It's a tough sell. And probably you have to find ways to exaggerate the threat. . . . [P]art of me sympathizes with the case for disingenuousness. . . . I sometimes wish that we could have, over the next five or ten years, a lot of horrid things happening—you know, like tornadoes in the Midwest and so forth—that would get people very concerned about climate change. But I don't think that's going to happen."[10] From the opening ceremony of the Copenhagen climate negotiations to Al Gore's documentary, *An Inconvenient Truth*, to public comments from leading climate scientists, to the echo chambers of the blogosphere, fear and alarm have been central to advocacy for action on climate change.

For those advocating action in this manner, the resulting dynamic is a positive feedback loop. Efforts are made to convince people of the urgency and importance of climate change by explaining the science. When people do not respond as advocates would like, the intensity of the campaign is ratcheted to a new level, with ever-more alarming warnings put forward in an effort to get the desired response from the public. Ultimately, this strategy collapses under its own weight, because science does not compel action, and arguments that cannot be well supported by science will be found out. For instance, at an annual meeting of Nobel Prize winners held in London in 2009, the Nobelists issued a statement in which they compared the threat of climate change to the threat of thermonuclear war.[11] A commentator from the BBC found this a bit over the top: "While debates around climate change are still qualified by the words 'might,' 'could' and 'predicted,' it's probably fair to say that the average person in the street may view the comparison of carbon emissions with things that can vaporize a major city in seconds as unhelpfully alarmist and perhaps just a little bit silly."[12]

In some respects, the campaign to convince people that climate change is a threat may have been too successful, such that people have come to believe things that the science cannot support. For instance, a 2007 *New York Times*/CBS poll found that of the three-quarters of people who believed that weather over the past few years had been stranger than normal, 43 percent attributed that weather to "global warming" and a further 15 percent to "pollution/damage to the environment." Yet, as most scientists will explain, weather events and even climate patterns over a period of years simply cannot be attributed to greenhouse gas emissions. Detecting changes in climate requires decades of observations. A very cold winter or two does not disprove a decades-long warming trend, and a series of damaging hurricanes is not evidence of a human influence.

Some advocates, including some scientists, seek to have things both ways when they assert that a particular weather event is "consistent with" predictions of human-caused climate change. The snowy period of early 2010 along the U.S. East Coast saw those opposed to action suggesting that the record snow and cold cast doubt on the science of human-caused climate change, while at the same time those calling for action explained that the weather was "consistent with" the forecasts from climate models. Both lines of argument were misleading. Any and all weather is "consistent with" predictions from climate models under a human influence on the climate system. Similarly, any and all weather is also "consistent with" failing predictions of long-term climate change. Simply put, weather is not climate. It takes decades and longer to detect and attribute a human signal on the climate system. Given the degree of politicization of the climate debate, we should not be surprised that even the weather gets politicized.

By the same token, it should come as no surprise that many in the public hold views about climate science that are way out in front of the scientific consensus on climate change as represented by the reports of the IPCC. The result is that when people learn what the science actually says, there is a risk that they will learn that their views are in fact incompatible with what the science can support, leading to a belief that the science has been overstated in public debate. This dynamic may help to

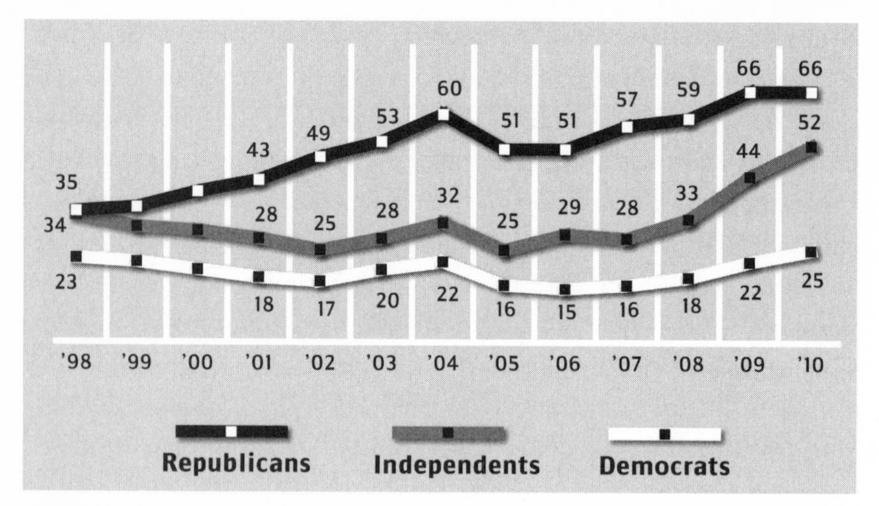

FIGURE 8.1 Percentage saying news of global warming is exaggerated, by party ID. Source: Gallup.

explain the trends seen in Figure 8.1, which shows an increasing number of the U.S. public saying that news of global warming is exaggerated, with the fastest rate of increase among Democrats and independents since 2006.[13] Other factors that could influence this trend might include the continued questions about aspects of climate science raised by opponents to action, who are more than willing to wage politics through science, and the vagaries of short-term weather, such as the paucity of hurricanes from 2006 through 2009 and several cold winters and cool summers in the same span. Whatever factors explain the trend, public trust in climate science appears to be a causality of the climate debate.

Efforts to use fear and alarm to motivate action are counterproductive to efforts to enact climate policies. Ted Nordhaus and Michael Shellenberger suggest that such efforts reinforce the partisan nature of the climate debate: "The louder and more alarmed climate advocates become in these efforts, the more they polarize the issue, driving away a conservative or moderate for every liberal they recruit to the cause." As the issue becomes polarized in partisan terms, it should come as no surprise that public opinion on the issue comes to resemble public opinion on most highly partisan issues.

Many scientists recognize the perils of leading with alarm. Vicki Pope, a British climate scientist, warned that "linking climate change to the latest extreme weather event or apocalyptic prediction" might get attention for scientists, but at a cost to the public perception of climate science.[14] But the temptation to grab attention has a powerful allure. About a month after Pope issued her warning to her colleagues, the organization that she leads was rebuked by a colleague for announcing publicly that "the impacts of climate change on the Amazon are much worse than we thought."[15] The work that was referred to had not yet been peer reviewed and was part of a much broader literature on the topic that suggested a range of different views and uncertainties. Yadvinder Malhi of the University of Oxford took journalists to task for reporting the story, and, by extension, Pope's organization as well. Highlighting "only the most catastrophic scenarios," he said with understatement, "has the potential to backfire."[16]

In fact, employing the fear factor seems to have already backfired. The data show that increasing numbers of the public have become more skeptical of the claims of climate scientists, believing that the science has been exaggerated. Nevertheless, support for action remains strong. Taken together, these data suggest that invoking fear and alarm serves most to undermine perceptions of science and offers precious little benefit to motivating action.

Demanding Certainty and Delivering It

Another manner in which the politicization of climate science manifests itself is when scientists ascribe more certainty to their work than is warranted. They do this, in part, in response to demands for certainty from politicians. For instance, in March 2009 Prime Minister Anders Fogh Rasmussen of Denmark explained to the scientific community at a major climate science meeting what he needed from them leading up to the Copenhagen climate conference later that year:

> But understand me correctly; at the end of the day, here in Copenhagen, we have—as politicians—to make the final decision, and to decide on exact figures, I hope. And this is a reason why I would give you the piece of advice, not to provide us with too many moving

> targets, because it is already a very, very complicated process. And I need your assistance to push this process in the right direction, and in that respect, I need fixed targets and certain figures, and not too many considerations on uncertainty and risk and things like that.[17]

Everyone would love knowledge to be certain and uncontroversial, of course. But one reason in particular that politicians demand certainty from scientists has to do with political accountability—it serves to transfer some of the basis for decision making from the public official to the expert. If things go awry, the politician can always point to the experts to deflect responsibility and say, "They told me to take these actions!"

Some scientists are rightly wary of demands for certainty when certainty is not forthcoming. Judy Curry, a climate scientist at Georgia Tech, worries about the consequences: "Politicians say, 'We need to reduce the uncertainty,' and I think that's contributed to a certain mindset where [climate scientists] try to reduce the uncertainty" when they talk about their research. "I'm a little bit worried about that political pressure," she said.[18]

Demands for certainty, however, don't just come from politicians. Climate scientists also impose such demands on themselves, in order to make their scariest projections even scarier. This leads to more problems. In one of the more widely quoted comments ever made by a climate scientist, Steve Schneider wrote:

> On the one hand, as scientists we are ethically bound to the scientific method, in effect promising to tell the truth, the whole truth, and nothing but—which means that we must include all the doubts, the caveats, the ifs, ands, and buts. On the other hand, we are not just scientists but human beings as well. And like most people we'd like to see the world a better place, which in this context translates into our working to reduce the risk of potentially disastrous climatic change. To do that we need to get some broad-based support, to capture the public's imagination. That, of course, entails getting loads of media coverage. So we have to offer up scary scenarios,

> make simplified, dramatic statements, and make little mention of any doubts we might have. This "double ethical bind" we frequently find ourselves in cannot be solved by any formula. Each of us has to decide what the right balance is between being effective and being honest. I hope that means being both.[19]

For his part, Schneider emphasized that this double ethical bind should never be resolved by resorting to mischaracterizing uncertainties. In response to the frequent use of his quote to suggest a green light for alarmism, Schneider wrote, "Not only do I disapprove of the 'ends justify the means' philosophy of which I am accused, but, in fact have actively campaigned against it in myriad speeches and writings."[20] Indeed, the vast majority of climate scientists that I have had the pleasure to get to know and work with over the years shares Schneider's passion for accurately conveying climate science to the public and placing it into its policy context. However, not all of their colleagues share this passion, coloring views of all of the climate-science enterprise.

Playing Politics Through Science

The politicization of climate science revealed in the East Anglia e-mails—as well as the appeals to fear and the efforts to misrepresent uncertainty—is far more common in the behavior of leading climate scientists than many in the community would care to admit. Most scientists might prefer to stay out of the public eye and simply conduct their research, but influential and activist leaders of the community have sought to achieve political outcomes—typically support for specific action in response to climate change—through science, especially by using and shaping science as a tool to defeat their political enemies.

Some of this we have explored already. In Chapter 2 we saw Stephen Pacala explain that the "stabilization wedges" paper was written to undercut any claims by President George W. Bush or his administration that technological innovation would be necessary to reach emissions-reduction goals, in an effort to counter the work of New York University's Martin Hoffert and colleagues. Pacala explained that the paper succeeded in the political objective far beyond expectations. Unfortunately, although Pacala won the short-term political battle, as Chapter 2 also shows, the stabilization wedges paper has been shown to be misleading and to dramatically understate the technological effort needed to accelerate decarbonization of the global economy. Pacala's political victory set back climate policy.

In Chapter 5 we saw Nobel laureate Paul Crutzen explain that his paper extolling the possibilities of injecting aerosols into the stratosphere was meant not to justify this type of geoengineering but instead to scare politicians into accepting more conventional forms of climate mitigation. Crutzen's efforts, however, backfired, as many interpreted his work as *legitimizing* that sort of radical geoengineering as a possible alternative to mitigation.

Pacala's and Crutzen's efforts to influence politics through science are far from unique. In 2006 journalist Elizabeth Kolbert described why Stanford's Ken Caldeira coined the term "ocean acidification" to describe the effects of carbon dioxide being absorbed into the ocean: "for its shock value."[21] Michael Mann of Pennsylvania State University explained that a desire to influence public opinion prompted him and a coauthor to incorrectly characterize IPCC projections as predictions in a popular book on the IPCC results. "The purists among my colleagues," he said, "would rightly point out that the potential future climate changes we describe, are, technically speaking, projections rather than predictions. . . . In this case, use of the more technically 'correct' term is actually less likely to convey the key implications to a lay audience."[22] In this case, the desire for public impact trumped scientific accuracy.

Hans von Storch, Nico Stehr, and Dennis Bray characterized the general attitude of many climate scientists: "The concern for the 'good' and 'just' case of avoiding further dangerous human interference with the climate system has created a peculiar self-censorship among many climate scientists. Judgments of solid scientific findings are often not made with respect to their immanent quality but on the basis of their alleged or real potential as a weapon by 'skeptics' in a struggle for dominance in public and policy discourse."[23]

In my own work over the years I've seen firsthand the desire to wage political battles through science. Several years ago the editor of a leading scientific journal asked me to dampen the message of an article of mine—which had been both peer reviewed and accepted for publication—for fear that it would be seized upon by those seeking to defend their interests in business-as-usual energy policies. In 2001 I was invited by the National Academy of Sciences to participate in a briefing of senators and the U.S. secretary of the treasury on climate change. Upon seeing my name listed as a participant in this high-level affair, several colleagues contacted me to suggest that I should downplay a core finding of my research on disasters—which we saw in Chapter 7—that future climate impacts depend much more on growing human vulnerability to climate than on projected changes in the climate itself, even under the most pessimistic assumptions of the IPCC. One colleague wrote to me at the time, "I think we have a professional (or moral?) obligation to be very careful what we say and how we say it when the stakes are so high."[24] The message being sent to me was clear.

Of course, leading climate scientists have not been alone in playing climate politics through science. Some opposed to action on climate change, rather than highlighting uncertainty, offer a competing set of certainties in the debate, claiming that the human impact is negligible or even beneficial; thus, no actions are needed. Others take a different approach, relying on uncertainties to justify inaction. These opponents have, as Gavin Schmidt worried, blown out of proportion papers at odds with the views of most other scientists. As I argued in Chapters 1 and 2, uncertainty, far from justifying inaction, should lead us to ask instead what actions might be warranted. Uncertainty by itself does not compel action, as some, relying on the so-called precautionary principle, might argue, but it does force us to clarify that which we value and to make choices accordingly.

Ironically, those who are opposed to action—who are variously dismissed as skeptics, deniers, or other pejorative terms—share a core assumption in common with most of those who are calling for action, especially their opponents in climate science. Specifically, both camps agree that climate science should be the ground on which battles over climate politics are to be waged. The presence of climate skeptics in the debate reinforces the belief that the path to action must go through a shared vision of climate science, its ground rules, accepted practices, and the implications that it has for action or inaction. As a result, if climate skeptics did not exist, they would have to be invented.

Regardless of the content of any particular politicized scientific argument, the form is generally the same. Dan Sarewitz, director of the Consortium for Science, Policy, and Outcomes at Arizona State University, diagnoses the failed formula, which holds regardless of point of view: "If only, the complaint goes, those (a) conservatives (b) liberals (c) environmentalists (d) industrialists or (e) ignorant members of the public would understand the facts, or stop manipulating the facts for their own political gain, we could arrive at rational solutions to the problems we face."[25] The operating assumption here seems to be that if my political opponents understood the facts of the matter exactly as I do, then they would come to share my values and political preferences. As seductive as this assumption is, it is a flawed analysis of how political action occurs for several reasons.

One reason, as Sarewitz points out, is that science is particularly ill-suited to resolving political disputes. Science is a diverse enterprise of methods and disciplines, and consequently there will always be a diversity of legitimate views on complex subjects. Advocates for this or that action will always be able to selectively pick and choose among findings to support whatever course of action they prefer. Sarewitz argues that rising political stakes can often lead to increasing scientific uncertainties, at least in the short term, as more perspectives and approaches are brought to bear upon particular questions.

Another reason that arguing politics through science is a poor approach to dealing with complex political issues is that such an approach has a disproportionately negative downside for scientific institutions and the process of science. Advocates for inaction on climate change who base their arguments in science (however flawed these arguments may be), being in the minority, can play David to the majority's Goliath. If David is caught exaggerating or simply making a mistake, the damage to science is small because not much is expected of an underdog, especially one repeatedly characterized as being out of the mainstream. But when those who present themselves as representing science itself are caught, the damage can be dramatic. After all, if the "outsider" lies or makes a mistake, who cares? He can be (and typically is) dismissed as just a kook, hence the recourse to the term "denier" in the climate debate, first coined to evoke comparisons with Holocaust deniers. But if advocates who claim to represent the mainstream scientific establishment tell lies or make mistakes, they become an argument in favor of inaction. There is thus an asymmetry in the consequences of politicizing science that falls in the favor of those opposed to action. As Mike Hulme writes, "Climate scientists, knowingly or not, become proxies for political battles. The consequence is that science, as a form of open and critical enquiry, deteriorates while the more appropriate forums for ideological battles are ignored."[26]

Consider what happened in early 2010 when it was widely recognized that the IPCC had made a series of egregious errors in its 2007 report on the melting of Himalayan glaciers. The report claimed that the glaciers could be fully gone by 2035, which when published led to headlines around the world. It turns out that the basis for the claim was an offhand comment made by an Indian scientist to a reporter in 1999 with no basis in the peer-reviewed literature. The magnitude of the error was compounded because its first reports were harshly dismissed by the chair of the IPCC as "voodoo science" and the stuff of "climate change deniers and school boy science."[27] The mistake also revealed a significant deviation in procedures by the IPCC for including information in its report. The fact that the IPCC had made a mistake was less important than that the IPCC failed to live up to its own standards and its chairman loudly derided those complaining about the error.

Even worse, however, was that the mistake may not have been an innocent one. According to a report in the *Daily Mail*, the lead author of the relevant chapter in the IPCC report explained that the error resulted from an effort to influence political outcomes on climate change: "[The 2035 prediction] related to several countries in this region and their water sources. We thought that if we can highlight it, it will impact policy-makers and politicians and encourage them to take some concrete action. It had importance for the region, so we thought we should put it in." Soon after this was reported the IPCC author told a blogger that he never said such things, to which the *Daily Mail* reporter vehemently disagreed. Like much in the climate debate we are left with he said/he said.[28] If the original report is accurate in this case, then the IPCC, which is supposed to review science in the context of being "policy neutral," apparently evaluated the glacier claim not by its scientific accuracy or support in the literature, but instead by its potential to influence political outcomes. Whether that was the case in this circumstance or not, there is ample evidence that the IPCC may have more systemic issues along these lines, which likely helps to explain the egregious misrepresentation of disasters and climate change documented in Chapter 7.

For better or worse, climate change as a political debate has been fought through science, with the scientific community taking a leading role in advocating for action. At times this advocacy is in the open, such as that conducted in recent years by NASA's James Hansen, who characterizes his role in the climate debate using religious terminology. But too often the advocacy performed by scientists is stealthy, hidden behind the asserted authority and impartiality of climate science. Such stealth advocacy is doomed to backfire when hidden agendas are revealed and when efforts to manipulate science and scientific assessment processes become known.

Perhaps the most frustrating aspect of efforts to wage climate politics through climate science is that the battle for public opinion has essentially been won, as indicated by the analysis presented in Chapter 2. Public opinion on climate change has seen its ups and downs over the past decade or so, but it has been remarkable for its stability. In broad terms, a majority of people accept the idea that humans do have an influence on the climate system and are supportive of action on climate policy. This level of support is sufficient for policy action to take place. The challenge is thus one of policy design, not rallying public opinion. Consider that in 2009 in the United States, Democrats—who, generally speaking, favor action on climate change—had control of the White House and both chambers of Congress, yet were unable to pass meaningful climate legislation. Similarly, across the world where there is far more support for action on climate policies, action to decarbonize the global economy has been minimal at best, as detailed in Chapter 4. These policy shortfalls are not the result of insufficient public will, and will not be addressed by waging public battles over climate science. Rather, they are policy failures in the face of sufficient, and in many cases ample, public support for action.

In the end, fighting political battles under the guise of debating climate science will likely have little impact on overall public support for action, but rather, on public confidence in climate science itself. As the depth of the politicization of climate science has become more fully visible, the consequences for the scientific enterprise have been significant. An irony here is that such efforts were always doomed to fail, as waging a political battle through science confers a significant advantage to those who are presented as being outside the scientific mainstream. David and Goliath are held to different standards, no matter which is supposed to be the good guy.

There is also a more subtle, and probably more significant, impact of waging climate politics through climate science: the debate serves to prevent alternative views from emerging in the policy debate. The debate over the science reinforces a Manichaean view of climate politics and fosters a with-us-or-against-us mentality. Those offering alternative visions of climate policy—whether they start by accepting certain aspects of the science or not—are unwelcome. Steve Rayner of the University of Oxford railed against the closed nature of the debate in 2005 in terms of support for the Kyoto Protocol under the UN's Climate Convention: "Unfortunately, support for Kyoto has become a litmus test for determining those who take the threat of climate change seriously. But, between Kyoto's supporters and those who scoff at the dangers of leaving greenhouse gas emissions unchecked, there has been a tiny minority of commentators and analysts convinced of the urgency of the problem while remaining profoundly skeptical of the proposed solution. Their voices have largely gone unheard."[29]

For his efforts, Rayner was placed on a list of 400 "climate skeptics" by Senator James Inhofe, a Republican from Oklahoma vehemently opposed to action on climate change. Not surprisingly, some advocates for action have used the list to delegitimize Rayner's views in discussions of climate policy. Similarly, when Chris Landsea, my collaborator on hurricane-loss work and a widely recognized expert on hurricanes, argued in 2005 that it was premature to assert a causal connection between greenhouse gas emissions and hurricane behavior, Kevin Trenberth, lead author for the relevant IPCC chapter on hurricanes, warned ominously that "politics is very strong in what is going on" and without a hint of irony dismissed Landsea's arguments using a common strategy: "He is linked to the skeptics."[30] Landsea, who resigned from the IPCC over concern about how his views were being handled, has since then seen his stance vindicated as the consensus of the community in a paper published in early 2010 by a team of researchers who had previously been at odds on the issue.

Matthew Nisbet of American University offers an explanation as to why climate scientists are susceptible to politicizing their science:

> Scientists are also susceptible to the biases of their own political ideology, which surveys show leans heavily liberal. Ideology shapes how scientists evaluate policy options as well as their interpretations of who or what is to blame for policy failures. Given a liberal outlook and strong environmental values, it must be difficult for scientists to understand why so many Americans have reservations about complex policies that impose costs on consumers without offering clearly defined benefits. Compounding matters, scientists, like the rest of us, tend to gravitate toward like-minded sources in the media. Given their background, they focus on screeds from liberal commentators which reinforce a false sense of a "war" against the scientific community.[31]

Nisbet's thesis is backed up by polling data on partisanship among scientists. A 2009 poll found that only 4 percent of researchers in the geosciences (a category that includes most climate scientists) self-identified themselves as Republicans and 62 percent as Democrats. This split contrasts dramatically with the 23 percent of the public at large who self-identify as Republicans and the 35 percent as Democrats.[32]

One of the most prominent liberal commentators, Paul Krugman, Princeton economist and columnist for the *New York Times*, announced a general strategy to delegitimize alternative points of view in 2008: "The only way we're going to get action [on climate change], I'd suggest, is if those who stand in the way of action come to be perceived as not just wrong but immoral."[33] Krugman's call to arms has been heard by activist bloggers, including many scientists who seek to close down any semblance of debate on issues of climate change, often invoking the authority of science to do so. For instance, one senior climate scientist from the University of Illinois took issue with the views that Andy Revkin referenced on his popular blog at the *New York Times*. He threatened Revkin with ostracism if he continued to discuss unpopular views: "Your reportage is very worrisome to most climate scientists. Of course, your blog is your blog. But, I sense that you are about to experience the 'Big Cutoff' from those of us who believe we can no longer trust you, me included."[34]

Such threats make it seem that some of those advocating action on climate change are willfully creating a zone of policy ignorance by not only ignoring those who offer different policy views but also castigating those who hold such views as immoral. This is typically done through arguments about science, but as indicated by Australia's prime minister, a distinction between debates over science and those over political action is typically conflated. Unfortunately, if the lack of progress on climate policy results not from a failure of political will, as I've argued, but a failure of policy design, then we need more options for action at the table, not fewer. This means opening up the debate to new voices and views, even (and perhaps especially) those who offer uncomfortable or challenging views. Nevertheless, opening up the debate is exactly what advocates for action have been trying to avoid, and the authority of science has been the main tool of exclusion. This sort of politicization, which is subtle and far from the cartoonish images of scientists falsifying data popular in some circles, serves to undermine climate policy and reinforce a status quo characterized by rancor and gridlock.

Dan Sarewitz summarizes the general problem with waging climate politics through science: "Contrary to all our modern instincts, then, political progress on climate change requires not more scientific input into politics, but less. Value disputes that are hidden behind the scientific claims and counterclaims need to be flushed out and brought into the sunlight of democratic deliberation. Until that happens, the political system will remain in gridlock, and everyone will be convinced that they are on the side of truth."[35] Fixing climate policy depends in no small part on fixing the role of climate science in the debate.

Fixing Climate Science by Putting Advocacy in Its Proper Place

When Al Gore testified before the United States Congress in 2007, he used an analogy to describe the challenge of climate change: "If your baby has a fever, you go to the doctor. If the doctor says you need to intervene here, you don't say, 'Well, I read a science fiction novel that told me it's not a problem.' If the crib's on fire, you don't speculate that the baby is flame retardant. You take action."[36] With this example Al Gore was not only advocating a particular course of action on climate change but also describing the relationship between science (and expertise more generally) and decision making. In Gore's analogy, the baby's parents (i.e., in his words, "you") are largely irrelevant to the process of decision making, as the doctor's recommendation is accepted without question. The expert is the source of wisdom that determines the course of policy action.

But anyone who has had to take their child to a doctor for a serious health problem or an injury knows that the interaction among patient, parent, and doctor can take a number of different forms. In my book *The Honest Broker: Making Sense of Science in Policy and Politics*, I describe various ways that an expert (e.g., a doctor) might interact with a decision maker (e.g., a parent) in ways that lead to desirable outcomes (e.g., a healthy child). Experts have choices in how they relate to decision makers, and these choices have important effects on decisions but also the perception and role of experts in society. Fixing climate science will require a more sophisticated approach to the relationship of experts and decision makers than the simple idea that scientists call for action and policy makers respond.

Gore's analogy provides a useful point of departure to illustrate the four different roles for experts in decision making that are discussed in *The Honest Broker*. The four categories are very much ideal types; the real world is more complicated, but nonetheless I do argue that they help to clarify roles and responsibilities that might be taken by experts seeking to inform decision making:

The Pure Scientist seeks to focus only on facts and has no interaction with the decision maker. The doctor might publish a study that shows that aspirin is an effective medicine to reduce fevers. That study would be available to you in the scientific literature.

The Science Arbiter answers specific factual questions posed by the decision maker. You might ask the doctor what are the benefits and risks associated with ibuprofen versus acetaminophen as treatments for fever in children.

The Issue Advocate seeks to reduce the scope of choice available to the decision maker. The doctor might hand you a packet of a medicine with orders to give it to your child. The doctor could do this for many reasons.

The Honest Broker of Policy Options seeks to expand, or at least clarify, the scope of choice available to the decision maker. In this instance the doctor might explain to you that different actions are available, from wait-and-see to taking different medicines, each with a range of possible consequences.

Scholars who study science and decision making have long appreciated that efforts to focus experts only on the facts, and to keep values at bay, are highly problematic in practice. As Sheila Jasanoff of Harvard University has written about science in politics generally: "The notion that scientific advisors can or do limit themselves to addressing purely scientific issues, in particular, seems fundamentally misconceived." Sarewitz echoes this theme in the context of climate change, explaining that because science is not well suited to resolving political disputes, politics finds a home in debates putatively about science: "As good children of the Enlightenment, we should turn to science to establish the facts about problems such as climate change before deciding what policies to implement. Yet the types of things that scientists are good at figuring out don't have much to do with the types of things that politicians need to decide."[37]

How science and political values overlap in practice depends on the context of decision making, which is located in a broader societal context of values, institutions, culture, interests, uncertainties, and many other factors. For instance, consider the Pure Scientist or Science Arbiter described above. How would you view your doctor's advice to take ibuprofen if you learned that she had received $50,000 last year from a large company that sells ibuprofen? Or upon hearing advice to perhaps forgo medicine for this particular ailment, what if you learned that she happened to be an active member of a religious organization that promoted treating sick children without medicines? Or if you learned that her compensation was a function of the amount of drugs that she prescribed? Or perhaps the doctor was receiving small presents from an attractive drug-industry representative who stopped by the doctor's office once a week? There are countless ways in which extrascientific factors can play a role in influencing expert advice. When such factors are present they can lead to stealth issue advocacy, which I define as efforts to reduce the scope of choice under the guise of focusing only on purely scientific or technical advice. Stealth issue advocacy has great potential for eating away at the legitimacy and authority of expert advice, and even for the corruption of expert advice.

Being aware of potential corruption does not tell us, however, what forms of advice make sense in what contexts. In *The Honest Broker* I argue that a healthy system of decision making will benefit from the presence of all four types of advice (with the same actor even taking on different roles in different situations), but also that, depending on the particular context of a specific decision, some forms of advice may be more effective and legitimate than others. Specifically, I suggest that the roles of Pure Scientist and Science Arbiter make the most sense when values are broadly shared and scientific uncertainty is manageable (if not reducible). An expert would act as a Science Arbiter when seeking to provide a response to specific questions posed by decision makers and as a Pure Scientist if no such guidance is given. (In reality, however, the Pure Scientist may exist more as myth than anything else.)

In situations in which there are conflicting values or when scientific certainty is contested—that is to say, in almost every political issue involving scientific or technical considerations—then the roles of Issue Advocate and Honest Broker of Policy Options are most appropriate. The choice between the two would depend on whether the expert wants to reduce or expand the available scope of choice. Stealth issue advocacy occurs when one seeks to reduce the scope of choice available to decision makers but couches those actions in terms of serving as a Pure Scientist or Science Arbiter (e.g., "The science tells us that we must act . . .").

So your child is sick and you take him to the doctor. How might the doctor best serve the parent's decisions about the child? The answer depends on the context.

If you feel that you can gain the necessary expertise to make an informed decision, you might consult peer-reviewed medical journals (or a medical Web site) to understand treatment options for your child instead of directly interacting with a doctor.

If you are well informed about your child's condition and there is time to act, you might engage in a back-and-forth exchange with the doctor, asking her questions about the condition and the effects of different treatments.

If your child is deathly ill and action is needed immediately, you might ask the doctor to make whatever decisions are deemed necessary to save your child's life, without including you in the decision-making process.

If there is a range of treatments available with different possible outcomes, you might ask the doctor to spell out the entire range of treatment options and their likely consequences to inform your decision.

Even in the superficially simple scenario of a doctor, a parent, and a child, it's clear that the issues are complicated. Understanding the different forms of this relationship is the first step toward the effective governance of expertise.

Perhaps not surprisingly, the policy response to climate change has neglected the complexity of the relationship of experts and decision makers. That relationship is dominated by issue advocacy—and far too often, stealth issue advocacy, where political battles are hidden behind science. Notable for their absence are institutions that serve as honest brokers of policy alternatives. Improving science advice to decision makers on climate change depends in no small part on bringing political disputes out into the open and creating a safe place for discussions of uncertainties and a wide range of policy options. This will require a new sort of leadership among climate scientists and, in particular, a willingness for some to step back from the overt and stealth advocacy that has come to consume the climate science community, and for leading scientific organizations to take on a role as an honest broker of policy alternatives. Climate policy is notable for its lack of such honest brokers.

We have choices in how experts relate to decision makers. These choices shape the ability to use expert advice well in particular situations, but also shape the legitimacy, authority, and sustainability of expertise itself. Whether we are taking our children to the doctor, seeking to use military intelligence in a decision to go to war, or using interdisciplinary climate research to inform decarbonization and adaptation policies, better decisions will be more likely if we pay attention to the role of expertise in decision making and the different forms that it can take.

CHAPTER 9

Obliquity, Innovation, and a Pragmatic Future for Climate Policy

IMAGINE THAT NATIONS around the world were to decide that they want to increase human life spans. They might start their work by deciding on a goal. Today, the global average life span is about sixty-nine years. One nation might suggest a target of a seventy-five-year average life span by 2050. Others might criticize the lack of ambition in this target, and press for eighty-five years by 2050. A great amount of time could be spent arguing about what the right target should be, and the negotiations might founder there, without even an agreement on the goal, let alone implementing it.

But let's say that a target is agreed upon. One way to reach it would be to put an economic price on death: countries would be responsible for paying some amount when someone dies. There could be a death tax or, even more creatively, a cap on the number of allowable deaths in each country, with a decreasing number of permits issued each year. If a country experiences more deaths than they have permits for, then they will have to buy permits from other countries that are experiencing a lower death rate, and thus have surplus permits. Advocates for this approach would claim that the cap would stimulate innovation in medical research and prompt behavioral changes that lead to longer life expectancies. The permit market would mean that the market would decide efficiently how resources are best allocated, so that the "lowest hanging fruit"—the simplest or lowest-cost actions to increase life expectancy—would get funded. The legal cap would provide certainty in achieving the targeted extension of life expectancies.

Such a scheme would not be without its challenges. For instance, some countries, such as Japan, already have long average life spans, and thus have less ability to make progress. They might argue for some sort of excess permits to reflect their historical achievements on health. Similarly, many countries in sub-Saharan Africa have shorter average life spans, reflecting historical realities and long-standing inequities. These countries might argue for some time period to develop before being included under any sort of binding agreement, in recognition of history and their present circumstances.

Perhaps, a very creative policy entrepreneur might argue, a system of "death offsets" might be created, under which rich countries could get additional "death permits" by paying poor countries to improve their health outcomes, where there is much greater progress to be made. With a declining cap on deaths and a market-based instrument to efficiently allocate death permits, the world would inevitably move toward increasing the average human life span. Who could be against such an elegant approach to improving health outcomes? No one, obviously, except perhaps the nefarious industry-funded "health deniers"!

I have used this parable in many talks over the past few years, and inevitably the audience laughs at the notion of a cap-and-trade regime for extending life expectancies. The laughter is appropriate, as the example is a joke. But it is no more of a joke than efforts to reduce emissions of carbon dioxide employing the same type of policy. In much the same way that the amount of carbon dioxide in the atmosphere is a consequence of our actions, average human life spans are a consequence of actions, and consequences are extremely difficult to modulate directly via policy.

Policies are more effective when they focus on causes, not consequences. Despite the lack of internationally negotiated goals, the world has in fact seen a steady increase in average human life spans. This increase has been achieved with a focus on causes of poor health. We advanced average human life spans not by targeting it directly, but by focusing on a large set of individual diseases and public health challenges that lead to mortality. When we make progress on these health challenges, the fruits of our efforts show up in statistics of advancing average human life spans. The best route to advancing human life spans is indirect, focused not on seeking to change it directly, but on the phenomena that influence it. This logic is so compelling that any policy maker suggesting a global cap-and-trade regime on death would not be taken seriously, and for good reason.

Global emissions of carbon dioxide are the result of economic activity and the technologies of energy production and consumption. People in countries around the world expect to see continued economic growth, which means that, all else being equal, emissions will increase. In Chapter 2 I call this contemporary reality (which is grounded in a deeply held global and ideological commitment to economic growth) the iron law of climate policy. The iron law holds that for the foreseeable future, efforts to reduce emissions through a willful contraction of economic activity are simply not in the cards. Countries around the world—rich and poor, North and South—have expressed a commitment to sustaining economic growth, and these commitments are not going to change anytime soon, no matter how much activists, idealists, or dreamers complain to the contrary. People will pay some amount for environmental goals, but only so much before drawing the line. That is just the way it is, regardless of whether economic growth measures what matters most to a country's well-being or if there are other metrics that might better capture quality of life.

Given the iron law, then, there are only two ways that decarbonization of economic activity will occur. One is through improving the energy efficiency of the economy—which includes changes both in the efficiency of specific activities, such as steelmaking and automobile gas mileage, as well as in the nature of the economy, such as the increasing role played by less energy-intensive services sectors—and the other is through decarbonization of the energy supply.

In recent decades improvements in energy efficiency have been the primary driver behind decarbonization of the global economy, with decarbonization of energy supply a distant second. This must change, if decarbonization is to accelerate to rates consistent with low targets for atmospheric carbon dioxide, such as 350, 450, or even 500 ppm. The need to focus on decarbonizing the energy supply results from the simple fact that under all scenarios of future energy demand, even those based on relatively modest economic growth, the world will need a vastly larger energy supply. Whether that need is 30 percent, 50 percent, or 100 percent more than today's supply does not alter the basic calculus that to reach low stabilization targets implies a massive transition to a nearly carbon-free global energy supply. Advances in efficiency can alter the pace at which that transition takes place, but they will not change the end point. There is good news in that advances in efficiency often make sense for reasons other than decarbonization, most important being economic reasons, and thus have an independent justification. The bad news, of course, is that to the extent that efficiency gains do result in economic benefits they can stimulate the economy, leading to greater energy demand and thus greater emissions, unless the energy supply itself is decarbonized. Scholars and practitioners have debated the degree to which energy-efficiency improvements lead to greater energy demand. Even if there is no positive feedback effect on energy demand from energy-efficiency improvements, decarbonization of the energy supply still must happen.

But how do we accelerate decarbonization of the global economy through increasing the deployment of a low- or zero-carbon energy supply? Here again, we can look to efforts to improve health outcomes. Human life span is addressed disease by disease, public health issue by public health issue, and it will be through a similarly focused approach that decarbonization is accelerated. Of course, innovation in medicine and public health does not guarantee better health outcomes, just as innovation in military technology does not guarantee victory in battle. Both health care systems and military conflicts reflect deep complexities in society and its institutions that are independent of technological innovation. Similarly, success in energy innovation is necessary but not sufficient to decarbonize the global economy.

To make progress accelerating the decarbonization of the global economy we are going to have to overcome two further aspects of conventional wisdom that underlie the debate over climate change, and ultimately take a new approach. The first bit of conventional wisdom that we must overturn is the idea that we use too much energy. As described in Chapters 3 and 4, the world will need much more energy in the future, and this is reflected in all scenarios of leading national and international energy agencies. There are presently 1.5 billion people around the world who lack access to a basic energy supply. This is almost five times the population of the United States and three times the population of the EU. If all of these people were instantaneously granted energy access and produced the global average per capita emissions of 4.4 metric tons of carbon dioxide, they would add more than 20 percent to global carbon dioxide emissions from the consumption of fossil fuels, and the emissions from the bottom 193 countries would instantaneously double.[1] The uncomfortable reality of large increases in emissions that accompany greatly expanded access to energy may help to explain why some scenarios for stabilizing concentrations of carbon dioxide leave the vast majority of those lacking access in the dark. But as we'll soon see, providing energy to those without access may be a key element of an "oblique" strategy to accelerate decarbonization.

The second aspect of conventional wisdom that we must overturn is the idea that fossil fuels are too cheap. A fixture of the climate debate has been the notion that the best way to make alternative energy sources more cost competitive with fossil fuels, and thus preferable in markets of energy supply, is to increase the costs of fossil fuels. However, this line of thinking, whatever its merits to be found in economic theory, runs smack into the unyielding iron law of climate policy. Evidence from around the world clearly shows that while people are willing to accept some increases in their energy costs, this willingness has its limits. And those limits are nowhere near those needed to motivate the complete decarbonization of the world's energy supply. It is on these shoals that climate policy has repeatedly foundered.

If substantially raising the price of fossil fuels is not a viable option, then we must think differently about the challenge. British economist John Kay provides some insight as to how differently our thinking must be for complex challenges: "If you want to go in one direction, the best route may involve going in the other. Paradoxical as it sounds, goals are more likely to be achieved when pursued indirectly. . . . Oblique approaches are most effective in difficult terrain, or where outcomes depend on interactions with other people. . . . Obliquity is characteristic of systems that are complex, imperfectly understood, and change their nature as we engage with them."[2] An oblique approach to climate policy is necessitated by the complexity and inherent "wickedness" of the issue, one that, as Mike Hulme has argued, is not a problem to be solved but a condition that can be managed for better or worse.

An oblique approach to decarbonization begins with the realization that whatever the future holds for energy demand, we can stabilize concentrations of carbon dioxide at a low level only via a massive expansion of the availability of the carbon-free energy supply. This means that any policies based on the idea that we, as a world, use too much energy are on the wrong track from the start. The world needs more energy. Much more. Over the long term increasing demand for energy and limits on the easily accessible supply of fossil fuels will together lead to some degree of decarbonization of economic activity, though how much and how fast are uncertain. But as we've seen, the historical rate of so-called spontaneous decarbonization—and even a rate twice as fast—is inadequate to stabilize carbon dioxide concentrations at a low level, such as 450 ppm.

Justifying dramatically accelerating decarbonization of the global economy requires an oblique approach involving the application of policy jujitsu. Consider the 1.5 billion people who presently lack access to energy. It seems obvious that basic access to energy is, if not a human right, a matter of human dignity. A main obstacle to securing access to energy for all is that present supply options, which include both fossil fuels and their alternatives, cost too much. Cheaper sources of energy would advance human dignity by making it easier to secure energy access for all; the cheaper that energy is, the broader the benefits and the greater the demand for energy. Meeting this massive demand requires technological innovation in support of diversification of the energy supply. Greater diversification implies decarbonization of the energy supply and economic activity. The greater the rate of innovation, the greater the potential pace of decarbonization.

Expanding access to cheap and secure sources of energy appeals to overlapping interests. The advantages for countries with large numbers of people without access to a basic energy supply are clear. Others around the world would benefit from the opening of new markets, both through building the infrastructure of the energy supply and through the explosion of economic activity in these newly powered communities, creating new opportunities for trade and growth. A focus on expanding the energy supply thus unites two virtues—addressing fundamental human needs while providing inescapable motivation for accelerating decarbonization of the global economy—and eliminates the pathological trade-off between development and emissions reductions implicit (and at times explicit) in conventional climate policies. An oblique focus on expanding the energy supply may offer the best prospect of accelerating decarbonization. Heading in what might seem to be the wrong direction—dramatically expanding the energy supply, access, and consumption—may offer the best prospect for getting us to where we should be headed.

Of course, if, through some discovery or technological advancement, the costs of fossil fuels fall dramatically in coming years, then a policy goal of dramatically expanding access to energy could be achieved without accelerating decarbonization of the energy supply. The recent progress in freeing up natural gas from shale deposits suggests that there are still surprises in the technologies of fossil fuels. Nevertheless, even with such discoveries it seems unlikely that the costs of fossil fuels will decline significantly; enormous effort and subsidies have been devoted to providing fossil fuel energy at the lowest costs possible (so low, in fact, that some governments tax them due to people's willingness to accept paying a bit more). It is unlikely that the costs of fossil fuels can be further reduced, meaning that they will still cost too much for many people to use, and diversification will remain a compelling strategy for securing a reliable energy supply.

In fact, the costs of fossil fuels should increase in coming decades for simple reasons of supply and demand, providing a separate, compelling basis for innovation-focused energy policies. All of this implies a policy challenge to develop alternative sources of energy supply that are cheaper than fossil fuels. Google.org has expressed this policy challenge elegantly in terms of a simple equation: RE < C, which stands for *R*enewable *E*nergy that *costs less* than *C*oal.[3] Google.org explains that "today renewable electricity costs too much to compete with coal."[4] Developing cheaper electricity will aid in expanding energy access around the world and would also contribute to energy security and economic development. A significant side benefit to such a strategy is that, if successful, it will also contribute to the dramatic acceleration of the decarbonization of the global economy. Google.org's innovative approach reflects the notion that rather than being too cheap, fossil fuels cost too much.

Economics alone probably won't be enough to motivate a dramatic reduction in consumption of fossil fuels, even if alternatives to fossil fuels are developed at lower costs. The reason for this lies in the fact that the process of securing lower-cost alternatives to fossil fuels may have the perverse effect of motivating an accelerated extraction of fossil fuels as owners of those resources see no better time than the present to capitalize on their value, as the resources will only be worth less in the future. German economist Hans Werner-Sinn calls this the "green paradox."[5] What this means is that some form of political commitment to leaving fossil fuels in the ground will likely have to accompany innovation of alternative sources of energy. Such a commitment will be made far more likely with alternatives cheaper than fossil fuels available, and will be impossible without those alternatives. Innovation thus has to be front and center.

Critical to driving this innovation will be government. Government-sponsored work has been integral to other efforts, such as the fight against disease, where investments have focused on a disease-by-disease approach. Likewise, progress on energy innovation will occur technology by technology, via investments in innovation. Fortunately, as progress in medicine as well as other fields such as agriculture and defense attests, innovation is something that governments at national and international levels have often managed well over periods of decades.

In 2009 Arizona State University and the Clean Air Task Force, a nonprofit organization based in Boston, organized a series of workshops examining the role of the public sector in fostering technological innovation in climate policy.[6] Their resulting analysis focused on efforts in the United States, but their conclusions have implications for accelerating energy innovation worldwide:

To improve government performance, and expand innovation options and pathways, Congress and the administration must foster competition within the government. Competition means allocating resources based on performance. The United States, they argue, relies too much on a single agency for energy innovation. Agencies or programs that show results should be rewarded with additional resources; those that do not should be downsized or terminated. The implication is that energy innovation needs to be conducted in a diverse set of institutions, with performance continuously being evaluated.

To advance greenhouse gas–reducing technologies that lack a market rationale, the government should selectively pursue energy-climate innovation using a public-works model. If it is in the public interest to expand the secure, low-cost energy supply and in the process to slow or reduce the emission of greenhouse gases, then there is a public-sector role for investing in energy technologies that complement a more globally focused effort to secure energy access for all. The workshop participants offer vaccines for pandemic flu, flood-control dams, and aircraft carriers as analogous technologies supported by the public.

To stimulate commercialization, policy makers must recognize the crucial role of demonstration projects in energy-climate innovation, especially for technologies with potential applications in the electric-utility industry. Demonstration projects can help to overcome concerns about technical and cost uncertainties that often limit the adoption of new technologies. They point to carbon capture and storage as an example of such a technology that would benefit from full-scale demonstration in order to prove (or disprove) its viability.

To catalyze and accelerate innovation, the government should become a major consumer of innovative energy-technology products and systems. Governments spend a large amount of money on energy and as such can stimulate market demand, driving down prices and building confidence in products that are close to being market ready.

There is ample experience worldwide with public and private efforts to stimulate innovation to serve as the basis for designing new approaches to energy innovation, building on existing institutions. Such experience includes that of the Consultative Group on International Agricultural Research,[7] the Global Alliance for Vaccines and Immunization,[8] and the accumulated experience of the world's military establishments which have overseen innovation in many areas of technology.

We seem to know how to do innovation well enough to get started on the challenge. That leaves two questions: How will we carry it out, and how we will pay for it?

It is important to understand that the technologies of the future are almost certainly going to be technologies of the present, only better. Breakthroughs that lead to fundamentally new sources of energy—such as nuclear fusion—while possible, are unlikely. The alternatives to fossil fuels are well known and include various technologies of wind, solar, biomass, nuclear, hydropower, carbon neutral fossil fuels, and a few others. The technologies of consumption involve issues of storage of energy, such as in batteries that might be used to displace liquid fuels for transport, and the management of energy such as in advanced electrical grids capable of handling intermittent sources. All of these technologies are with us in one form or another, but not many are developed to the point where the technology or economics is suggestive of readiness for large-scale deployment.

Consequently, any successful innovation-based approach to decarbonization will benefit from a policy stance that might be called technological agnosticism, since we do not presently know where advances might lie. A broad portfolio of technologies and practices should be supported in a global energy-innovation policy, despite the fact that no one energy technology will be universally popular. Every energy technology has its supporters and opponents, but to give in to every naysayer risks turning NIMBY (not in my backyard) into BANANA (build absolutely nothing, anywhere, near anyone). Under democratic systems of governance, tough choices will inevitably be made involving deployment of energy technologies. That should not preclude innovation, lest we limit our options before those options are even available. The uncomfortable reality is that the more technologies deemed politically unacceptable, the greater the challenge of accelerating decarbonization, the longer we'll depend upon fossil fuels, and the longer more than a billion people will lack basic access to energy.

It is often said that public investments in energy technologies should not "pick winners." This stance is only partially correct. One important reason for public investment is in fact to pick winners—those technologies that are cheaper than fossil fuels and can reliably supply copious amounts of new energy in the decades to come. The important thing is not to pick specific winners in advance, which includes taking options off the table before they have been explored.

Paying for this research is the other major challenge facing policy makers and the public. After all, experience in medicine and the military suggests that innovation does not come cheap. Consider, for instance, that the United States government alone invests about $30 billion annually in medical research and development and about $80 billion in military research and development. Such investments have been made for a period of many decades, complemented (and sometimes exceeded) by private-sector investment. Successful innovation in the energy sector that contributes to the sustained acceleration of decarbonization of the world's economy will require similar levels of investment for similarly long time periods.

Where will that money come from? Isabel Galiana and Chris Green of McGill University have argued for an innovation-led approach to climate policy, and they estimate that a $5 per metric ton tax on carbon

TABLE 9.1 Implications of a $5 per metric ton carbon tax for the prices of coal, oil, gasoline, and natural gas

	Carbon-tax equivalents		
	CO_2 (tonnes)	*at $5/t CO_2 ($)*	*2009 price ($) per unit*
Metric ton of coal	2.86	14.30	16–110
Barrel of oil	0.37	1.85	45–70
Gallon of gasoline	0.0088	0.044	2.00–2.50
1,000 cubic feet of natural gas	0.055	0.22	10–11
1,000 cubic meters of natural gas	2.025	8.10	~400

Source: I. Galiana and C. Green, "An Analysis of a Technology-Led Climate Policy as a Response to Climate Change," Copenhagen Consensus on Climate, 2009. http://fixtheclimate.com/fileadmin/templates/page/scripts/downloadpdf.php?file=/uploads/tx_templavoila/AP_Technology_Galiana_Green_v.6.0.pdf.

dioxide would have a relatively small effect on fossil fuel prices, with the exception of the cheapest forms of coal, and would raise as much as $150 billion per year—about $30 billion in the United States and China, a bit less in Europe, and substantial amounts elsewhere.[9] Table 9.1 shows their estimates for the effects of such a tax on the price of fossil fuels. A tax of $5 per metric ton is also attractive because a tax at this level has been proposed by the CEO of ExxonMobil, the energy company most often identified as the chief villain opposed to action on climate change.[10] If ExxonMobil can support such a tax, then it would suggest a lot of possible room for a broader political consensus. Galiana and Green compare a low carbon tax to the fuel tax placed on gasoline by the Eisenhower administration in support of building the U.S. interstate highway system. That tax had public support because it was linked to investments that paid off in tangible near-term benefits.

It is important to emphasize that the point of a carbon tax at this level is not to change people's behavior, to restrict economic activity, or to price fossil fuels at a level higher than alternatives. The purpose of a low carbon tax is to raise revenues for investments in innovation. Galiana and Green suggest that a key part of an innovation-focused approach to accelerating decarbonization would be a commitment to a forward-looking increase in the carbon tax, perhaps a doubling over ten years. Such a commitment would signal a manageable increase in the price of fossil energy and raise additional revenues along the way.

Of course, if innovation actually begins to result in an accelerated decarbonization of the economy, then it will prove politically easier to raise the tax, as its effects on consumers would be increasingly less profound as the economy slowly decouples economic growth from emissions. Over time, as the tax increases, there would be a convergence in outcomes between those who have argued for starting with a high price on carbon and the view expressed here of starting with a low tax. To some degree, a rising tax would help to address the so-called green paradox.

Such a tax would most efficiently be set "upstream," on the act of extracting fossil fuels itself. The precise amount of the tax itself—whether $5 per metric ton, or $10, or only $3—is less important than that the tax be implemented at the highest price politically possible. Using politics as the metric for pricing the tax would be far superior to trying to meet some theoretical ideal, such as through estimates of the "social cost of carbon" developed through complex economic models that require discerning trends and preferences decades and longer into the future. Starting with a low tax seems like an obvious strategy in any case for the simple reason that a high price on carbon—whether via a tax or through trading regimes—is just not going to be implemented in the near term. The only way to a high carbon tax is to start low.

The Indian government seems to have already recognized the importance of an innovation-led approach based on funding raised from low prices on fossil fuels. In February 2010 Pranab Mukherjee, India's minister of finance, proposed a 50 rupee (about $1) tax on every metric ton of coal mined in or imported by India to support a "National Clean Energy Fund." The tax would raise about $600 million per year to be invested in alternative energy with a focus on reducing pollution from coal energy and also expanding energy access.[11] The Indian government's plan provides an example of the approach recommended here, including a focus on expanding energy access as a basis for gaining short-term political buy-in to the tax.

At the international level, there would be many details to work out, of course, but there is no doubt that this type of scheme would be far less complicated than that of the present climate regime. Imagine what a difference it would be for nations to gather at an international climate convention with a single point to negotiate: at what level do we wish to collectively tax carbon dioxide emissions from fossil fuels? With a decision reached on that question—which still would not be easy to achieve—nations could then turn to the question of how to invest the proceeds in a manner that addresses political demands, but also does not compromise on the need to focus the revenues on energy innovation in an efficient and effective manner.[12] One of the greatest challenges facing this approach would be the ability of nations, individually and collectively, to direct funds raised by a carbon tax to energy innovation. Given the debt situation facing many nations around the world, not least the United States, it would be exceedingly tempting to agree to a carbon tax and then funnel the proceeds into general government revenue, giving innovation short shrift.

Once one takes a look at the simple mathematics of emissions reduction described in Chapters 3 and 4, it is inescapable that the world does not in any practical sense have "all the technology that it needs" to achieve low levels of stabilization of carbon dioxide. This point is further reinforced when a commitment is made to providing access to energy presently lacked by 1.5 billion people around the world. Energy innovation must be at the core of any policy focused on stabilizing carbon dioxide levels in the atmosphere.

We have seen the simple mathematics that show convincingly that progress on accelerating decarbonization of the global economy will be a consequence of technological innovation. Thus, whatever policies are ultimately adopted, the real measure of progress will begin with the effects of those policies on the rate of innovation. With little attention to innovation, progress will be correspondingly small. If there is a single variable that will serve as a measure of progress toward emissions reduction or carbon-intensity goals, it will be the proportion of global energy consumption that comes from carbon-neutral (or even-negative) sources. In recent years that value has been well under 10 percent of total global consumption.[13] To achieve stabilization at low levels, that proportion will eventually have to exceed 90 percent. The mathematics of decarbonization are as simple as the practical challenge is difficult.

For decarbonization, the most effective path is to focus on innovation funded by a low carbon price that rises over time. An innovation-focused approach retains some of the elements of a pricing scheme in that it prices carbon, albeit at a level far lower than followers of conventional wisdom demand. But the iron law of climate policy limits what can be done in the near term. By explicitly connecting carbon pricing with energy innovation, a virtuous circle is enabled that allows those asked to pay the tax to see its benefits and thus builds the support necessary to sustain investments over decades and longer. Ironically, a more direct focus on decarbonization of the global economy means a less direct focus on climate change, as there are other reasons to accelerate decarbonization focused on expanding energy access, increasing security, and reducing costs. As we've seen with the oil catastrophe in the Gulf of Mexico in 2010, less reliance on fossil fuels might lead to environmental benefits that go well beyond climate.

An important and perhaps uncomfortable aspect of the approach recommended here is that it leaves much uncertain: we do not know, exactly, how we are going to achieve the long-term goals of providing energy access, security, and low cost that, if achieved, will also drive accelerated decarbonization of the global economy. This uncertainty is inescapable, but not crippling. Rather, this sort of uncertainty exists in other challenges that we collectively face. How will we improve human life spans over coming decades? How will we secure economic stability? How will we maintain peace? No one pretends that there is a comprehensive solution, much less a definitive road map to success, for any of those questions. Rather, we proceed incrementally on many parallel paths while learning from, and adjusting to, experience. Acceptance of policy uncertainty can be liberating as well: no one depends on a cost-benefit analysis integrated over the better part of a century to assess individual policies focuses on health, security, or the economy.

To summarize, the approach to accelerating decarbonization recommended here has four related elements. First, it begins by setting goals. Policy making is more likely to succeed with a clear sense of direction. The world should aim to provide secure access to inexpensive energy to everyone on the planet by some date in the not too distant future, perhaps 2030 or 2040. Achieving that goal will require unprecedented innovation in energy technologies such that alternatives to fossil fuels are cheaper than fossil fuels. Climate thus becomes relegated to a secondary consideration. John Kerry, Democratic senator from Massachusetts, explained a very similar logic when discussing reworked climate legislation in early 2010 after cap and trade was determined not to be the way forward, "It's primarily a jobs bill, and an energy independence bill and a pollution reduction–health–clean air bill. Climate sort of follows. It's on for the ride."[14]

Second, achieving that goal will necessitate rapid innovation in energy technologies. Innovation will be necessary in both efficiency of energy use and technologies of energy production. Efficiency gains can lead to reduced costs of energy for particular users, but may also have the consequence of increasing overall demand for energy. Thus, progress will also have to be made in developing sources of energy that are cheaper than fossil fuels. Innovation policies should be implemented based on the lessons of experience from areas such as health, agriculture, and the military. Progress in innovation policies will be enhanced by adopting an attitude of technological agnosticism from the outset, where the broader goals are set and the details are not prejudged.

Third, raise funds to invest in innovation via a low carbon tax, priced as high as politically possible, perhaps $5 per metric ton and applied "upstream" where fossil fuels are removed from the earth. The point of the tax would not be to change behavior, as any price on energy high enough to be felt will indeed change behavior—most likely the voting behavior of citizens protesting against costly energy. A commitment to a long-term increase in the tax, rising at a level that keeps pace with energy-technology innovation but does not violate the iron law of climate policy, will provide the market a forward-looking price signal. If the approach recommended here were to succeed, then decades hence the world will have a high carbon tax, widespread deployment of low-carbon technologies, and a decarbonized global economy.

Fourth, progress should be continuously monitored, and policies should be adjusted based on performance. Key variables will include the number of people remaining without access to electricity and the cost of energy. If the world makes progress in providing vast amounts of energy at costs less than fossil fuels, one inevitable consequence will be an accelerated decarbonization of the global economy and lower emissions.

The practical need for such an approach can be seen in a decision by the World Bank in April 2010 to underwrite the building of a massive new coal plant in South Africa. The decision was opposed by environmentalists who called on the United States and European nations to block the loan, in order to halt construction. Pravin Gordhan, South Africa's finance minister, explained why the project was so important: "Today, the South African economy is two-thirds larger than it was in 1994, when Nelson Mandela took office as the country's first democratically elected president. With this growth has come strong new demand for electricity. Millions of previously marginalized South Africans are now on the grid. Unfortunately, as in other major emerging economies, supply has not kept pace." He also explained that the need for energy trumped concerns about carbon dioxide emissions: "We are using every tool at our disposal—legislative, regulatory and fiscal—to promote clean and renewable energy and manage demand. If there were any other way to meet our power needs as quickly or as affordably as our present circumstances demand, or on the required scale, we would obviously prefer technologies—wind, solar, hydropower, nuclear—that leave little or no carbon footprint. But we do not have that luxury if we are to meet our obligations both to our own people and to our broader region whose economic prospects are closely tied to our own."[15]

Gordhan's dilemma is a real-world reflection of the analysis presented in this book: "A question that has to be faced is whether stunting growth prospects in our region will in any way serve the goal we all share of eliminating greenhouse gas emissions over the long term. Whatever paths we take toward that goal, whether shifting to renewables and nuclear, or finding ways to keep harmful gases out of the atmosphere once created, the journey will inevitably be costly, requiring massive investments in technology, research and re-engineering the ways in which we live and do business. It will also require a true spirit of consensus and collaboration." When the time came to approve the World Bank loan to South Africa, the US, UK, and the Netherlands all chose to abstain, such that they were able to save face with domestic political constituencies but also not stand in the way of South Africa's development prospects. A political dilemma was thus resolved, but the decarbonization challenge persists.

The approach laid out above to focus decarbonization policy on expanding energy access via innovation in energy technology funded by a low carbon tax is but a rough outline, leaving many specifics to be worked out. The plan it offers, though, provides a much greater likelihood for success than the current policies, focused on targets and timetables for reducing emissions, which has been at the center of international negotiations for the past several decades. Still, this recommended approach could fail, too. Such an outcome is a risk of any climate policy.

The prospect of policy failure has led many observers of climate policy to suggest a need for a "Plan B" focused on geoengineering the earth's climate system. Chapter 5 argues that the idea of a simple technological fix is appealing, but illusory. To qualify as a technological fix a technology must meet three criteria, according to Sarewitz and Nelson: (1) the technology must largely embody the cause-effect relationship connecting problem to solution, (2) the effects of the technological fix must be assessable using relatively unambiguous or uncontroversial criteria, and (3) research and development is most likely to contribute decisively to solving a social problem when it focuses on improving a standardized technical core that already exists.

Technologies of geoengineering fail comprehensively with respect to the three criteria of a technological fix. Efforts to modulate the global earth system in an effort to counteract the consequences of increasing carbon dioxide in the atmosphere are certain to fail in the same manner that efforts to introduce cane toads into Australia as a tool of ecological management also failed. The global climate system is enormously complex and remains poorly understood in many respects. Efforts to intervene are just as likely to lead to undesirable outcomes as to beneficial ones. In any case, intervention in the climate system brings forth a whole host of legal and social challenges that have no simple answers. For instance, under a regime of engineering, who bears responsibility for climate extremes that take place, with uncertainties about the causal role of the geoengineering intervention? It seems clear that those who would seek to assert control of the climate then own the resulting weather outcomes, regardless of what the science may say about attribution.

Geoengineering is best left in the realm of the speculative. Even so, we should expect research related to geoengineering to continue to be conducted, for the simple fact that it is closely related to aspects of basic research on the climate system (such as what happens when a volcano erupts, sending large amounts of aerosols into the upper atmosphere?). In important respects, research on geoengineering shares characteristics with research on nuclear weapons. Some would argue that such research should be restricted, simply to take the possibility of deployment off the table. Others will argue for research to keep options open, or to understand technologies that others may deploy. Whatever course is taken, and it seems likely that some forms of geoengineering research will take place, such research should always be conducted in parallel with discussions of the social and policy implications of the technology, just as has been the case with respect to research on other contested technologies, such as on nuclear weapons.

The capture and storage of carbon—which is sometimes lumped in with geoengineering, although it is more accurately a technology of remediation—ought to be pursued. The air capture of carbon dioxide refers to chemical, biological, and geologic technologies that might contribute to drawing down and stabilizing atmospheric concentrations of carbon dioxide. While such technologies do not offer anything like a silver bullet, they do at least meet the three criteria for a successful technological fix. However climate policy evolves, several decades hence policy makers may wish to adopt a "brute-force approach" to stabilization if drawing down carbon dioxide becomes a much higher priority or to "mop up" the remaining carbon dioxide that was left unaddressed by more conventional mitigation policies. As a result, any portfolio focused on energy policy innovation with a pragmatic philosophy of "technological agnosticism" should include some research focused on developing technologies of air capture. One outcome is certain: if air capture is ignored, then little innovation should be expected with respect to these technologies.

As discussed in Chapter 1, carbon dioxide is far from being the only human influence on the climate system. Other greenhouse gases, such as nitrous oxide and methane, play important roles, as do more obscure trace gases such as sulfur hexafluoride and various chlorofluorocarbons. Soot, known as "black carbon," and other aerosols also influence the climate system by affecting cloud formation, precipitation, and the reflectivity of snow and ice. Patterns of land use, irrigation, fertilization, urbanization, and other human activities also have discernible influences on the climate. So, while many policy makers and advocates have emphasized carbon dioxide as the dominant or most important human influence on climate, it should be clear that carbon policy is not climate policy. The diversity of human influences on the climate system means that even if the world successfully decarbonizes the global economy to levels consistent with stabilization of carbon dioxide concentrations at low levels (such as 450 ppm), the world will still face a human-caused climate change problem.

To better align scientific understandings of the complexities of human influences on climate with policy, it is important to thus distinguish between carbon policy and climate policy. One important way to reflect this alignment would be to refocus the UN Framework Convention on Climate Change as a Framework Convention on Carbon Dioxide (and Other Long-Lived Greenhouse Gases). Such a focus would recognize that carbon is not all that matters with respect to the human influence on climate and emphasize the futility of trying to manage the enormously complex global climate system under a single policy instrument.

The degree to which the focus on carbon has become conflated with issues that best stand on their own merits was reflected in a research paper on whales published in early 2010. The author of that paper argued, "If you think about whales and fish in terms of their carbon, there is a potential for using carbon offset credits as an additional incentive for rebuilding this population."[16] According to the study, a single blue whale contains about 9.4 metric tons of carbon, equivalent to the per capita emissions of about eighteen months by a single American. The blue whale's "carbon value" would be only about $500 based on recent prices for carbon on market exchanges. To think that saving the whales might be tied to their role as a means of carbon storage is to take carbon policy to an utterly absurd end point. Saving whale populations makes good sense independent of their value as a carbon storage device. In similar fashion, placing other aspects of climate policy on the shoulders of carbon policy has led to a policy framework that has collapsed under its own weight.

The consequence of a narrowly focused carbon policy is a need for a more broadly conceived climate policy. As scholars have begun to point out, there are other policy instruments that might be used to address other human influences on climate, beyond carbon dioxide. The Montreal Protocol, which deals with the harmful effects of chlorofluorocarbons on the upper atmosphere, might be extended to address other influences of these chemicals on the climate system. Black carbon is not presently addressed under the climate convention but might be addressed under a separate policy instrument. In similar fashion policies focused on other aspects of the human influence on climate could be considered under existing and newly developed policy instruments. A diversified approach will require letting go of the fantasy that it is possible to deal with climate change comprehensively under a single, global policy instrument, and to resist the urge to allow narrow policies to become incrementally more complex. It also means that solutions to policy problems associated with forests, agriculture, biodiversity, global inequities, and other issues—even saving the whales—that have piggybacked onto the climate issue will once again have to stand on their own merits.

A reworked Climate Convention focused narrowly on carbon dioxide would need to reconsider its ultimate objective, which presently is focused on avoiding "dangerous interference" in the climate system. A more appropriate focus would be achieving decarbonization of the global economy to a level consistent with meeting long-term targets for stabilization of concentrations. A focus on decarbonization—perhaps obliquely as a consequence of the innovation necessary for expanding secure and low-cost energy access for those without—rather than danger would elevate the importance of technological innovation in carbon policy and de-emphasize the role of science, particularly long-term climate predictions. This would help to depoliticize climate science, as it would no longer serve as the fulcrum on which action is to be judged, and enable a healthier relationship of science and policy across the spectrum of issues related to a human influence on climate.

A focus on decarbonization rather than danger would have the beneficial effect of untethering adaptation from mitigation. The Climate Convention could do away with its narrow and scientifically untenable definition of "climate change" as referring only to those changes that result from accumulating greenhouse gases. Adaptation could then be freed from carbon policy and reunited with broader agendas focused on the development of sustainable, resilient communities able to weather extremes, disease, and other shocks irrespective of the precise human role in influencing the climate system. As we've seen, adaptation agendas can well stand on their own, with consideration of but independence from concerns about human influences on the climate system.

Focusing carbon policy on decarbonization and adaptation on development would place achieving goals of human dignity at the center of climate policies, which raises the odds of sustaining political support over the many decades necessary to make progress on these difficult challenges. By disaggregating climate policy into its component parts we can better align short-term costs with short-term benefits, which is critical to putting the policy jujitsu of oblique approaches to work. Besides that, placing human dignity and democratic ideals at the center of climate policies is also the right thing to do.

Acknowledgments

With hindsight, it seems that I've been working on this book for about twenty years or so. Over that time I have had the opportunity and often the distinct pleasure of working with and getting to know many physical and social scientists, policy makers, journalists, and others involved in the debate over climate change. I've benefited from the advice and wisdom of far more people than I can possibly acknowledge, but I am grateful to a few people in particular for their help over the years, and especially the intense period spent completing this volume.

At the top of the list, of course, are my parents, especially my father, Roger Pielke Sr., a brilliant and prolific scientist whose work I have no doubt will be received in very positive fashion by future generations of scientists and policy makers. Rad Byerly has been a mentor and friend for many years, and this book is deeply infused with his wisdom. Mickey Glantz shepherded me through the intense flames of the politicized world of climate science long before I could have survived such a journey myself. Over the years I have had the privilege of being able to work with and learn from some amazing people, including Dan Sarewitz, Bobbie Klein, Steve Rayner, Gwyn Prins, Eva Lövbrand, Mike Hulme, Mary Downton, Chris Green, Isabel Galiana, Nico Stehr, Hans von Storch, Suraje Dessai, Reiner Grundmann, Richard Tol, Max Boykoff, Tom Yulsman, Keith Kloor, Myanna Lahsen, Atte Korhola, Tom Wigley, Susan Avery, Andy Revkin, Stan Changnon, Laurens Bouwer, Marilyn Averill, Ryan Crompton, Kerry Emanuel, Peter Webster, Judy Curry, John McAneney, Joel Gratz, Jim White, Sam Fitch, Lisa Dilling, Rich Conant, Ben Hale, Peter Höppe, Eberhard Faust,

Elizabeth McNie, Jim White, Chris Landsea, Richard J. T. Klein, Linda Pendergrass, and Sarah Leshan. There are dozens of others not listed whose collaborations and collegiality have been much valued, as well as people whom I've never met but have found their work to be enlightening or inspirational. I have also benefited from the graduate students at the University of Colorado Environmental Studies program who have endured my courses focused on climate change and science policy over the past ten years. I've learned much from them.

The book was written at the Center for Science and Technology Policy Research at the University of Colorado–Boulder, within the Cooperative Institute for Research in Environmental Sciences (CIRES). One could not ask for a more fertile academic home to explore issues associated with the many dimensions of climate change. The depth of support provided by the center and its director, Bill Travis, and CIRES, through the tenures of Susan Avery and Koni Steffen, has been a remarkable asset and one that I value immensely. My academic work presented in the book spans the course of several decades and has benefited from academic research grants from numerous U.S. science agencies, most notably the National Science Foundation. I have also benefited from a close affiliation with the Breakthrough Institute, led by Michael Shellenberger and Ted Nordhaus, whose forward-thinking arguments have inspired much of my own thinking on energy policy in the broader context of aspirations for achieving human dignity.

A very special thanks goes to Ami Nacu-Schmidt for work over the past ten years that has been not only of remarkably high quality but utterly essential to the work that my colleagues and I do at the CIRES Policy Center. Her contributions to this volume are behind the scenes but very much appreciated.

In many respects, the arguments presented in this book have been forged in the fires of the Internet, where the notion of review—whether by peers or others—takes on an intensity quite unlike anything else that an academic experiences. I would like to thank the countless people who have commented in public or in private on ideas and arguments that I first aired on my blog. These ideas have been sharpened immensely by these exchanges, and I value the time invested by commenters, both critical and complimentary. If past is prologue, I expect such exchanges to continue upon publication of this book. You can follow those discussions and participate if you'd like at my blog: http://rogerpielkejr.blogspot.com. Of course, all responsibility for errors remaining in the book, muddled thinking, and imprecise presentation lies with me. If you do find an error, please let me know.

At Basic Books T. J. Kelleher helped me every step of the way from the first conceptualizations of the book to the details of individual arguments. I have learned a lot in the process of writing it. Thanks also to Sandra Beris for overseeing production and to copy editor Annette Wenda for offering much valuable advice in the final stages of writing. The book is better for their efforts.

Finally, last to acknowledge but first in all other respects is my family: Julie, Megan, Jacob, and Calvin.

Notes

Chapter 1

1. http://www.pbs.org/wgbh/pages/frontline/hotpolitics/etc/script.html.

2. Ibid.

3. S. Fred Singer, *Wall Street Journal*, August 1988, 22.

4. H. von Storch and N. Stehr, "Climate Change in Perspective: Our Concerns About Global Warming Have an Age-Old Resonance," *Nature* 405 (2000): 615.

5. Chris Russill, "Tipping Point Forewarnings in Climate Change Communication: Some Implications of an Emerging Trend," *Environmental Communication: A Journal of Nature and Culture* 2, no. 2 (2008): 133–153. Other nations trace the kickoff of their own domestic debates on climate change to other events, many during the 1980s.

6. M. Segal, R. A. Pielke, Jr., and Y. Ookouchi, "On Optimizing Solar Collectors Orientation Under Daily Nonrandom Cloudiness Conditions," *Journal of Solar Energy Engineering* 110 (1988): 346–348.

7. http://www.ucar.edu/communications/staffnotes/9806/here.html.

8. *Encyclopaedia Britannica* (1985), s.v. "atmospheric sciences."

9. Ibid.

10. In technical terms, understanding the challenge of stabilization requires understanding the dynamics of stocks and flows. See, for example, J. Sterman and L. Booth Sweeney, "Cloudy Skies: Assessing Public Understanding of Global Warming," *System Dynamics Review* 18 (2002): 207–240.

11. Research led by William Ruddiman suggests that humans have been influencing the climate for thousands of years. See *Plows, Plagues, and Petroleum* (2005; reprint, Princeton: Princeton University Press, 2010).

12. See, for example, I. G. Enting, T. M. L. Wigley, and M. Heimann, "Future Emissions and Concentrations of Carbon Dioxide: Key Ocean/Atmosphere/Land Analyses," *CSIRO Technical Paper* 31 (1994); and R. H. Socolow and S. H. Lam, "Good Enough Tools for Global Warming Policy Making," *Philosophical*

Transactions of the Royal Society, Ser. A (2006): 1–38. T. M. L. Wigley observes that the relationship of emissions and concentrations varies over time ("CO_2 Emissions: A Piece of the Pie," *Science* 316 [2007]: 829–830).

13. In 2007 the IPCC estimated for the period 2000–2005 that the oceans absorbed about 8.1 Gt of carbon dioxide per year, plus or minus 1.8 Gt. Land-surface processes resulted in a net uptake of about 3.3 Gt of carbon dioxide, plus or minus 0.6 Gt. The IPCC estimates the total annual natural sink of carbon dioxide for 2000–2005 to be 11.4 Gt, plus or minus 20.2 Gt (IPCC, "Working Group I: The Physical Science Basis of Climate Change," 2007, http://ipcc-wg1.ucar.edu/wg1/wg1-report.html).

14. Cf. C. L. Sabine et al., "The Oceanic Sink for Anthropogenic CO_2," *Science* 305 (2004): 367–371.

15. As referenced in Chapter 5, though, some think that the scale for sequestration via the land surface is virtually unlimited. Such proposals remain more science fiction than practical policy options.

16. http://www.latimes.com/news/local/la-me-logging25–2009sep25,0,5224449.story.

17. In 2007 the IPCC suggested that about 20 percent of annual emissions come from deforestation. A study in 2009 suggested a much lower value. See http://www.nature.com/ngeo/journal/v2/n11/abs/ngeo671.html.

18. Policy makers sometimes include an additional complication that involves converting other greenhouse gases into "equivalent" carbon dioxide terms and then referring to concentrations of this basket of greenhouse gases as "carbon dioxide equivalent." The analyses throughout this book focus on carbon dioxide. However, to get a sense of the implications of using a larger basket of greenhouse gases: 450 ppm of carbon dioxide–equivalent greenhouse gases includes about 400 ppm of carbon dioxide. At 388 ppm, the world is within a decade of this level at current rates of increase.

19. Chapter 5 will discuss in more depth the notion of such a threshold in the context of policy.

20. http://www.yaleclimatemediaforum.org/2009/08/g8s-2-degrees-goal/.

21. A doubling from 280 ppm is 560 ppm.

22. S. J. Tol Richard, "Europe's Long-Term Climate Target: A Critical Evaluation," *Energy Policy* 35 (2007): 424–432.

23. http://www.350.org/understanding-350.

24. http://www.350.org/messengers.

25. http://www.350.org/media/350countries.

26. http://www.yaleclimatemediaforum.org/2009/08/g8s-2-degrees-goal/.

27. http://www.nature.com/climate/2009/0910/full/climate.2009.95.html.

28. http://www.grida.no/publications/other/ipcc_tar/?src=/climate/ipcc_tar/wg1/268.htm.

29. http://www.ipcc.ch/pdf/assessment-report/ar4/wg1/ar4-wg1-chapter1.pdf, 96.

30. http://www.ipcc.ch/pdf/assessment-report/ar4/wg1/ar4-wg1-spm.pdf; http://www.sciam.com/article.cfm?id=the-future-of-climate-change-policy; http://www.supremecourtus.gov/opinions/06pdf/05–1120.pdf.

31. http://www.ipcc.ch/ipccreports/tar/vol4/index.php?idp=86.

32. Susan Solomon et al., "Irreversible Climate Change Due to Carbon Dioxide Emissions," *PNAS* 106, no. 6 (2009): 1704–1709, http://www.pnas.org/content/early/2009/01/28/0812721106.full.pdf+html; http://www.e360.yale.edu/content/feature.msp?id=2201.

33. Chapter 5 will discuss other reasons carbon dioxide became a focal point of climate policy.

34. Another important reason that the debate has centered on carbon dioxide can be gleaned from figure SPM.2 in the IPCC "Summary for Policy Makers" that shows the relative influence of various human factors on the climate system. Scientists use a metric called "radiative forcing" to quantify this influence. The figure shows the influence of carbon dioxide to be 1.66 watts per meter squared and the net human influence (after adding both positive and negative forcings) to be 1.6 watts per meter squared. The impression given, whether intended or not, is that the Earth's climate is more or less at a radiative-forcing equilibrium were it not for the influence of carbon dioxide (http://www.ipcc.ch/pdf/assessment-report/ar4/wg1/ar4-wg1-spm.pdf).

35. http://www.nature.com/news/2009/090729/full/news.2009.745.html.

36. http://www.wired.com/wiredscience/2009/07/jellyfish/.

37. R. A. Pielke, Sr., "Comment," *EOS: Transactions of the American Geophysical Union* (September 4, 2001): 394, 396.

38. R. A. Pielke, Sr., "Overlooked Issues in the U.S. National Climate and IPCC Assessments," *Climatic Change* 52 (2008): 1–11, available at http://www.climatesci.org/publications/pdf/R-225.pdf. In 2008 he also listed the influence of human-caused aerosols on clouds and precipitation and the regional radiative effect of atmospheric aerosols on atmospheric circulations and of aerosol deposition (soot, nitrogen) as being additional effects that have been either mostly ignored or insufficiently presented in the IPCC report (http://www.climatesci.org/publications/pdf/Testimony-written.pdf).

39. http://www.giss.nasa.gov/research/features/senate/page3.html.

40. http://www.eenews.net/climatewire/2009/07/14/archive/1?terms=james+hansen.

41. http://epa.gov/climatechange/endangerment/downloads/GHGEndangermentProposal.pdf.

42. http://www.guardian.co.uk/environment/2009/oct/04/climate-change-melting-himalayan-glaciers.

43. Mario Molina et al., "Tipping Elements in Earth Systems Special Feature: Reducing Abrupt Climate Change Risk Using the Montreal Protocol and Other Regulatory Actions to Complement Cuts in CO_2 Emissions," *PNAS* 106, no. 49 (2009): 20616–20621, http://www.pnas.org/content/early/2009/10/09/0902568106.full.pdf+html. See note 34 for a technical definition of "radiative forcing."

44. http://pielkeclimatesci.wordpress.com/files/2009/11/r-354.pdf.

45. Mike Hulme, *Why We Disagree About Climate Change* (Cambridge: Cambridge University Press, 2009).

46. http://www.ft.com/cms/s/0/38ec6d54-dee1–11de-adff-00144feab49a,dwp_uuid=2f2f2698-de6f-11de-89c2–00144feab49a.html.

47. http://dotearth.blogs.nytimes.com/2008/01/13/a-starting-point-for-productive-climate-discourse/.

48. For those seeking a more in-depth treatment of issues associated with carbon dioxide and greenhouse gases in particular, see the report of Working Group I of the IPCC. The primary scientific literature contains debates and much more in-depth discussions. Both the IPCC and the scientific literature can be daunting to those coming fresh to the issue. See http://www.ipcc.ch.

49. An example of such conventional wisdom can be found in various lectures of NASA's James Hansen. See, for instance, http://www.columbia.edu/~jeh1/2006/NAS_20060424.pdf.

50. M. M. Betsill and R. A. Pielke, Jr., "Blurring the Boundaries: Domestic and International Ozone Politics and Lessons for Climate Change," *International Environmental Affairs* 10, no. 3 (1998): 147–172; R. A. Pielke, Jr., and M. M. Betsill, "Policy for Science for Policy: Ozone Depletion and Acid Rain Revisited," *Research Policy* 26 (1997): 157–168.

51. Inadvertent Modification of the Stratosphere Task Force, *Fluorocarbons and the Environment: Report of Federal Task Force on IMOS Council on Environmental Quality* (Washington, DC: Federal Council for Science and Technology, June 1975).

52. Stephen O. Anderson and K. Madhava Sarma, *Protecting the Ozone Layer: The United Nations History* (London: Earthscan, 2002).

53. Thanks to Keith Stockton for insight into the case of acid rain.

54. See M. Smith, *Warnings: The True Story of How Science Tamed the Weather* (Austin: Greenleaf Book Group, 2010).

55. D. Sarewitz, R. A. Pielke, Jr., and R. Byerly, Jr., eds., *Prediction: Science, Decision Making, and the Future of Nature* (Washington, DC: Island Press, 2010); "Tomorrow Never Knows," *Nature* 463 (2010): 24.

56. Steve Schneider, "Global Warming: Neglecting the Complexities," *Scientific American* 286, no. 1 (January 2002): 62. See also "We're Altering the Environment Far Faster Than We Can Possibly Predict the Consequences," http://www.aarpmagazine.org/lifestyle/global_meltdown.html.

57. Sarewitz, Pielke, and Byerly, *Prediction.*

58. On this point see Kevin Trenberth, http://www.nature.com/climate/2010/1002/full/climate.2010.06.html.

59. http://www.fooledbyrandomness.com/cameronstatements.htm.

60. http://www.timesonline.co.uk/tol/news/environment/article7003622.ece.

61. Ibid.; "We're Altering the Environment."

Chapter 2

1. Chapter 5 discusses another key assumption that has hampered efforts to design and implement policies focused on adaptation: how we have framed the concept of "climate change."

2. http://www.timesonline.co.uk/tol/news/environment/article6658672.ece.

3. Polls use different phrasings and slightly different questions. Early examples include: 74 percent believe global warming is happening (1997); 79 percent believe global warming already does or will in the future have serious impacts (1999); 72, 75, and 74 percent "of those who have seen, heard, or read about global warming say that they believe in the theory that increased carbon dioxide and other gases will lead to global warming and an increase in average temperatures" (2000, 2001, 2002, respectively); 75 percent are somewhat or very concerned about global warming (2003).

4. http://woods.stanford.edu/research/majority-believe-global-warming.html.

5. http://www.gallup.com/poll/107569/ClimateChange-Views-Republican Democratic-Gaps-Expand.aspx.

6. http://www.gallup.com/poll/126560/Americans-Global-Warming-Concerns-Continue-Drop.aspx.

7. http://www.rasmussenreports.com/public_content/business/econ_survey_toplines/december_2009/toplines_climate_change_december_1_2_2009. In April 2010 an investigation at the request of the University of East Anglia found no evidence of any research misconduct. See http://www.uea.ac.uk/mac/comm/media/press/CRUstatements/Report+of+the+Science+Assessment+Panel.

8. The analysis presented here relies on poll data from the Gallup Polling organization collected over several decades. A systematic review of many such polls from many organizations over the same time period can be found in M. C. Nisbet and T. Myers's article "Twenty Years of Public Opinion about Global Warming" (*Public Opinion Quarterly* 71, no. 3 [2007]: 444–470), which presents data consistent with that presented here.

9. http://www.gallup.com/poll/106660/Little-Increase-Americans-Global-Warming-Worries.aspx.

10. An October 2009 poll from the *Wall Street Journal* and CBS News found that "the answers show a uniform, if moderate, decline in concern about climate

change in recent years—but very little change from a decade ago" (http://blogs.wsj.com/environmentalcapital/2009/10/28/poll-position-cap-and-trade-losing-support-nbcwsj-survey-finds/). Poll results are at http://online.wsj.com/public/resources/documents/wsjnbc-10272009.pdf.

11. For instance, an October 2009 poll by Pew showed a dip in public concern about climate change yet continued strong support for action (http://people-press.org/report/556/global-warming).

12. Nisbet and Myers, "Twenty Years of Public Opinion," 461.

13. Ibid.

14. http://online.wsj.com/article/SB10001424052748704107104574571613215771336.html.

15. http://www.gallup.com/poll/124595/Top-Emitting-Countries-Differ-Climate-Change-Threat.aspx.

16. http://siteresources.worldbank.org/INTWDR2010/Resources/Background-report.pdf.

17. http://www.gallup.com/poll/1633/Iraq.aspx#4; http://www.rasmussenreports.com/public_content/business/federal_bailout/september_2008/only_28_support_federal_bailout_plan.

18. P. Burstein, "Why Estimates of the Impact of Public Opinion on Public Policy Are Too High: Empirical and Theoretical Implications," *Social Forces* 84 (2006): 2273–2289. Compare Alan D. Monroe, "Public Opinion and Public Policy, 1980–1993," *Public Opinion Quarterly* 62 (1998): 6–28; and Benjamin I. Page and Robert Y. Shapiro, "Public Opinion and Public Policy," *American Political Science Review* 77 (1983): 175–190.

19. R. E. Dunlap, "Public Opinion and Environmental Policy," chap. 4 in *Environmental Politics and Policy*, edited by J. P. Lester, 2nd ed. (Durham: Duke University Press, 1997).

20. http://www.nytimes.com/1989/05/22/us/polls-show-contrasts-in-how-public-and-epa-view-environment.html?scp=2&sq=ozone+percent22public+opinion percent22&st=nyt.

21. Nisbet and Myers, "Twenty Years of Public Opinion," 460.

22. http://people-press.org/report/584/policy-priorities-2010.

23. http://people-press.org/report/485/economy-top-policy-priority.

24. The overriding importance of the economy is not a uniquely American phenomena. See http://www.harrisinteractive.com/news/ftharrispoll/hi_financial times_harrispoll_january2009_23.pdf.

25. http://www.e360.yale.edu/content/feature.msp?id=2210.

26. http://www.nature.com/climate/2010/1002/full/climate.2010.09.html.

27. Arguably, many in the public hold views of the directness and magnitude of human influence on the climate system that go well beyond what scientific research can support.

28. This conclusion holds even in the face of the dip in public support and trust in climate science observed in 2010.

29. http://www.e360.yale.edu/content/feature.msp?id=2210.

30. W. Lippmann, *Public Opinion* (New York: Harcourt, Brace, 1922), 197.

31. Thanks to Rafael Ramirez for this astute observation.

32. B. V. S. Girod, *Why Six Baseline Scenarios? A Research on the Reasons for the Growing Baseline Uncertainty of the IPCC Scenarios* (Zurich: ETH, 2006).

33. http://www.telegraph.co.uk/earth/earthnews/6248257/Planned-recession-could-avoid-catastrophic-climate-change.html.

34. http://www.guardian.co.uk/environment/2009/nov/29/rajendra-pachauri-climate-warning-copenhagen.

35. http://online.wsj.com/article/SB122904040307499791.html.

36. http://greeninc.blogs.nytimes.com/2009/01/13/chu-confirmation-update-answering-for-past-statements/.

37. http://corner.nationalreview.com/post/?q=MThjNDUzMWJlODg3ZTRkMDFhM2IzYzAxNmEyODA0OTU=.

38. http://dotearth.blogs.nytimes.com/2008/04/09/money-for-indias-ultra-mega-coal-plants-approved.

39. http://dotearth.blogs.nytimes.com/2008/04/09/money-for-indias-ultra-mega-coal-plants-approved/?apage=1#comment-14330.

40. http://thebreakthrough.org/blog/2008/04/maybe_horses_will_fly_developi.shtml.

41. Wind and solar will not fill the gap.

42. "Dark Days Ahead," *Economist*, August 6, 2009, 8.

43. http://www.guardian.co.uk/business/2008/jun/10/oil.france.

44. http://www.nytimes.com/2008/07/10/science/earth/10assess.html.

45. http://blog.algore.com/2009/06/whats_in_a_number.html.

46. http://online.wsj.com/article/SB124050061773748291.html.

47. http://www.capitolhillreports.com/033109.htm.

48. The actual text of the Ensign Amendment states that the intention is "to protect middle-income taxpayers from tax increases by providing a point of order against legislation that increase taxes on them, including taxes that arise, directly or indirectly, from Federal revenues derived from climate change or similar legislation." On his Web site Senator John Ensign (R-NV) suggested that "middle class" refers to citizens making $250,000 per year, a threshold he noted was frequently discussed by the Obama administration (http://www.senate.gov/legislative/LIS/roll_call_lists/roll_call_vote_cfm.cfm?congress=111&session=1&vote=00121).

49. Al Gore, *Our Choice: A Plan to Solve the Climate Crisis* (Emmaus, PA: Rodale Books, 2009).

50. http://www.abc.net.au/7.30/content/2009/s2700047.htm#.

51. S. Pacala and R. H. Socolow, "Stabilization Wedges: Solving the Climate Problem for the Next 50 Years with Current Technologies," *Science* 305 (2004): 968–972.

52. To convert from units of carbon to carbon dioxide one needs to multiply by 3.667. For instance, 3.667 tonnes of carbon dioxide contains 1 tonne of carbon. This section uses units of carbon for consistency with the analysis of Pacala and Socolow (ibid.). Details on carbon dioxide emissions and various concentration targets are discussed in depth in Chapter 1.

53. Ibid.

54. Hoffert's objections were chronicled by Elizabeth Kolbert in the *New Yorker* in March 2005 (http://www.newyorker.com/archive/2005/05/09/050509a_fact3). See also http://www.scientificamerican.com/article.cfm?id=plan-b-for-energy.

55. http://thebreakthrough.org/blog//2008/04/post_1-print.html.

56. http://homepage.mac.com/marty.hoffert/filechute/SocolowRedux.pdf.

57. http://thebreakthrough.org/blog//2008/04/post_1-print.html.

58. http://www.sciencemag.org/cgi/data/305/5686/968/DC1/1. Pacala and Socolow ("Stabilization Wedges") give a range of three to twelve wedges worth of emissions reductions due to the offsetting effects of the oceanic uptake.

59. The situation is even worse when you recognize that we do not yet have 60-mpg cars in automotive showrooms around the world and CCS technologies are not yet available.

60. http://www.eia.doe.gov/pub/international/iealf/tableh1co2.xls.

61. http://www.pbl.nl/en/publications/2009/Global-CO2-emissions-annual-increase-halves-in-2008.html.

62. http://www.eia.doe.gov/oiaf/ieo/excel/figure_81data.xls. The EIA projections also build in very aggressive assumptions of spontaneous decarbonization, but nonetheless run ahead of the stabilization wedge scenario.

63. http://sciencepolicy.colorado.edu/admin/publication_files/resource-2593–2008.08.pdf.

64. http://www.ipcc.ch/pdf/assessment-report/ar4/wg3/ar4-wg3-spm.pdf.

65. http://environmentalresearchweb.org/cws/article/futures/34156.

Chapter 3

1. In SI units 1 quad equals 1.055 exajoules, or 1,055,000,000,000,000,000 joules. One joule-second equals 1 watt.

2. http://www.aps.org/policy/reports/popa-reports/energy/units.cfm.

3. This can be thought of as a 1-GW power plant operating at 75 percent efficiency. However, since this is a rough calculation, such details are not necessary to elaborate precisely. How much is 750 MW in terms of residential electricity

consumption in the United States? "A 1,000 MW rated coal generator with a 75 percent capacity factor generates about 6.6 bn kWh [billion kilowatt-hours] in a year, equivalent to the amount of power consumed by about 900,000 homes in the Northeast but only 460,000 homes in the South. In other words, each MW of rated capacity for a coal plant in the Northeast generates the equivalent amount of electricity consumed by 900 homes in the Northeast but only about 460 homes in the South" (http://www.gasandoil.com/goc/features/fex32816.htm).

4. Some of this energy, of course, is liquid fuel. In Chapter 4, I use equivalent electricity generation as a metric of comparison for all energy consumption. Obviously, this is not directly substitutable with liquid fuels, but it does give some sense of magnitude, which is the point.

5. The total is (755–508) * 15 = 3,705. For 678 quads the total is (678–508) * 15 = 2,550. You easily can repeat this simple exercise with different assumptions. Recall that 508 quads is energy consumption estimated for 2010.

6. Vaclav Smil, *Energy at the Crossroads: Global Perspectives and Uncertainties* (Cambridge: MIT Press, 2003), 124.

7. See, for example, D. Botkin, "The Limits of Nuclear Power," *International Herald Tribune*, October 17, 2008; and K. Bradsher, "Earth-Friendly Elements, Mined Destructively," *New York Times*, December 26, 2009, A1.

8. http://www.washingtonpost.com/wp-dyn/content/article/2009/09/08/AR2009090804019.html.

9. http://www.iea.org/weo/electricity.asp.

10. http://www.guardian.co.uk/world/feedarticle/8767757.

11. http://www.iea.org/press/pressdetail.asp?PRESS_REL_ID=294.

12. http://www.nytimes.com/2009/09/20/world/europe/20denmark.html.

13. http://rogerpielkejr.blogspot.com/2009/12/how-large-is-global-energy-economy.html.

14. At twenty barrels of gasoline per barrel of oil this would equate to roughly a $0.15/gallon increase in the price of gasoline (http://tonto.eia.doe.gov/ask/gasoline_faqs.asp). According to the analysis of Isabel Galiana and Chris Green in 2009, this would be approximately equivalent to a $15/ton tax on carbon dioxide. See http://fixtheclimate.com/fileadmin/templates/page/scripts/downloadpdf.php?file=/uploads/tx_templavoila/AP_Technology_Galiana_Green_v.6.0.pdf.

15. ftp://ftp.fao.org/docrep/fao/012##876e/i0876e00.pdf.

16. ftp://ftp.fao.org/docrep/fao/012/i0876e/i0876e02.pdf.

17. http://www.washingtonpost.com/wp-dyn/content/article/2009/01/03/AR2009010301738.html.

18. http://www.msnbc.msn.com/id/28515983/.

19. http://www.foxnews.com/story/0,2933,477425,00.html?sub_secs=.

20. http://www.ft.com/cms/s/0/4e31fa7a-dc27-11dd-b07e-000077b07658.html.

21. Compare http://online.wsj.com/article/SB123621221496034823.html and http://www.newsweek.com/id/189293.

22. P. E. Waggoner and J. E. Ausubel, "A Framework for Sustainability Science: A Renovated IPAT Identity," *Proceedings of the National Academy of Sciences* 99 (2002): 7860–7865.

23. http://hdr.undp.org/en/reports/global/hdr2005/papers/HDR2005_Dikhanov_Yuri_8.pdf.

24. http://www-wds.worldbank.org/external/default/WDSContentServer/IW3P/IB/2008/09/02/000158349_20080902095754/Rendered/PDF/wps4620.pdf.

25. In technical terms the GDP used here is expressed in 1990 Gheary-Khamis U.S. dollars. See http://www.ggdc.net/maddison/Historical_Statistics/BackgroundHistoricalStatistics_09-2008.pdf.

26. http://www.eia.doe.gov/oiaf/ieo/excel/figure_81data.xls.

27. http://www.ipcc.ch/publications_and_data/ar4/wg3/en/contents.html.

28. For instance, J. E. Ausubel et al., "Carbon Dioxide Emissions in a Methane Economy," *Climatic Change* 12 (1988): 245–263.

29. R. A. Pielke, Jr., T. Wigley, and C. Green, "Dangerous Assumptions," *Nature* 452, no. 3 (2008): 531–532.

Chapter 4

1. http://www.eia.doe.gov/emeu/international/contents.html.

2. The use of a 2006 base year makes every argument in this chapter a bit understated, since emissions have increased since that time. Arguments suggesting that achieving decarbonization goals starting in 2006 is difficult are even stronger based on a 2011 start.

3. Carbon dioxide data are available at http://www.eia.doe.gov/pub/international/iealf/tableh1co2.xls. Data on GDP, converted to 1990 Gheary-Khamis dollars (to facilitate international comparisons), are available at http://www.ggdc.net/maddison/Historical_Statistics/vertical-file_09-2008.xls. The 1990 Gheary-Khamis dollars are the units used throughout and are explained at Maddison's Web site.

4. In fact, the effects of this reversal in China's decarbonization can be seen in the global data in Figure 3.7.

5. B. Pile, "We Have an Extremely Selfish Population," *Spiked*, February 25, 2009, http://www.spiked-online.com/index.php?/site/printable/6293/.

6. P. Marsh, "Make and Mend," *Financial Times*, February 9, 2009.

7. See Table 4.4 at the corresponding discussion end of the chapter for details on how to calculate nuclear power plant equivalents as a measure of the magnitude of the challenge of decarbonization.

8. The Dungeness B station produces 1090 MW of electricity, which at 75 percent efficiency is slightly more than 750 MW, the hypothetical "nuclear power station" equivalent unit used throughout the remainder of this chapter. See http://british-energy.com/pagetemplate.php?pid=91.

9. Committee on Climate Change, "Building a Low-Carbon Economy: The UK's Contribution to Tackling Climate Change," December 1, 2008, http://www.theccc.org.uk/reports/, 197; cf. K. Anderson, A. Bows, and S. Mander, "From Long-Term Targets to Cumulative Emissions Pathways: Reframing UK Climate Policy," *Energy Policy* 36 (2008): 3714–3722.

10. To put the debate over the Heathrow expansion into context, China is planning to build ninety-seven new airports by 2020. See http://www.france24.com/france24Public/en/archives/news/business/20080126-china-construction-airport-2020-soaring-demand-air-transport.php.

11. R. Harrabin, "UK's CO_2 Plan 'Certain to Fail,'" BBC News, February 11, 2009.

12. Interview with Japanese prime minister Taro Aso, *Financial Times*, June 10, 2009.

13. J. Kanter, "Tsunami of Criticism for Japan's CO_2 Goals," New York Times Green, Inc., 2009, http://greeninc.blogs.nytimes.com/2009/06/10/tsunami-of-criticism-for-japans-co2-goals/.

14. R. Maeda, "Japan Sees Extra Emission Cuts to 2020 Goal-Minister," Reuters, June 24, 2009, http://af.reuters.com/article/energyOilNews/idAFT19196720090624?sp= true.

15. V. Smil, "Light Behind the Fall: Japan's Electricity Consumption, the Environment, and Economic Growth," *Asia Pacific Journal: Japan Focus* (April 2007), http://www.japanfocus.org/-Vaclav_Smil/2394.

16. Cf. ibid.; and H. Geller et al., "Policies for Increasing Energy Efficiency: Thirty Years of Experience in OECD Countries," *Energy Policy* 34 (2006): 556–573.

17. The bullet points are quoted from the following article: http://www.eastasiaforum.org/2009/12/15/japan-and-forging-global-solidarity-at-copenhagen/.

18. http://wna-members.org/uploadedFiles/members/Info_Servies/Publications/PG_reactor08.pdf; see also Figure 4.11.

19. http://www.pm.gov.au/node/6006.

20. http://www.theage.com.au/news/national/rudd-takes-aim-at-the-us-over-warming/2007/12/12/1197135559287.html.

21. http://www.theage.com.au/news/national/rudd-withstands-conference-flak-to-appear-at-home-on-international-stage/2007/12/14/1197568264972.html.

22. http://www.bloomberg.com/apps/news?pid=20601081&sid=a3ipEehZOtLw&refer=australia.

23. http://www.news.com.au/couriermail/story/0,20797,23255892-953,00.html.

24. http://www.climatechange.gov.au/whitepaper/factsheets/pubs/031-australias-national-emissions-targets.pdf.

25. http://www.nytimes.com/2008/12/16/world/asia/16australia.html.

26. http://www.environment.gov.au/minister/wong/2009/pubs/mr20090504.pdf.

27. http://www.climatechange.gov.au/en/minister/wong/2009/media-releases/November/mr20091124.aspx.

28. http://www.smh.com.au/environment/australia-needs-herculean-efforts-to-meet-emissions-targets-20100212-nxmo.html.

29. These can be calculated from the data at: http://www.eia.doe.gov/oiaf/1605/ggrpt/excel/CO2_coeff.xls. In the analysis that follows I use 94.44 megametric tons of carbon dioxide per quad from coal, 70.00 for petroleum, and 53.06 for natural gas.

30. http://www.eia.doe.gov/pub/international/iealf/tableh1co2.xls; http://www.eia.doe.gov/emeu/cabs/Australia/Full.html.

31. http://www.world-nuclear.org/uploadedFiles/members/Info_Services/Publications/PG_reactor08.pdf.

32. http://www.aps.org/policy/reports/popa-reports/energy/units.cfm.

33. http://svc196.wic512d.server-web.com/cloncurry_solar_thermal_power_station.cfm.

34. http://www.eia.doe.gov/cneaf/alternate/page/renew_energy_consump/table1.html.

35. The wind turbines are provided by Shenyang Power Group. See http://online.wsj.com/article/SB125683832677216475.html.

36. http://www.eia.doe.gov/emeu/aer/pdf/pages/sec1_9.pdf.

37. The situation becomes even more complicated when issues of trade are considered in the accounting of emissions according to the location of their production or the consumption of goods that are produced. For instance, a considerable amount of China's national emissions are the result of economic activities resulting in export to other nations. See http://www.agu.org/pubs/crossref/2009/2008GL036540.shtml, which suggests that 9 percent of China's emissions are associated with goods shipped to the United States and 6 percent are associated with goods shipped to Europe.

38. http://www.bloomberg.com/apps/news?pid=20601091&sid=aWs0Pts2Kxes; http://economictimes.indiatimes.com/News/PoliticsNation/India-China-have-to-resist-pressure-on-climate-change-PM/articleshow/4765728.cms; http://www.guardian.co.uk/environment/2009/oct/01/india-us-climate-change.

39. http://www.nytimes.com/2009/12/10/business/economy/10consume.html.

40. http://www.reuters.com/article/idUSTRE5BA1ME20091211.

41. http://moef.nic.in/downloads/home/GHG-report.pdf.

42. http://rogerpielkejr.blogspot.com/2009/11/chinas-carbon-intensity-pledge.html.

43. http://www.nature.com/nature/journal/v462/n7270/full/462158d.html.

44. http://rogerpielkejr.blogspot.com/2009/10/two-views-of-bau-in-china.html.

45. South Africa, for example, has proposed emissions-reduction targets to 2020 that are no more realistic than those of other countries. See http://rogerpielkejr.blogspot.com/2010/01/magical-solutions-in-south-africa.html.

46. http://www.eea.europa.eu/pressroom/newsreleases/2009-greenhouse-inventory-report.

47. http://www.hs.fi/paakirjoitus/artikkeli/VieraskynpercentC3percentA4+EUn+lopetettava+ilmastokikkailu/1135249970758. Translation courtesy of Atte Korhola.

48. The EU here is defined as the EU-15. The final verdict is not yet in on the Kyoto Protocol's effects on EU decarbonization, as it runs through 2012.

49. http://www.cer.org.uk/articles/50_miliband.html.

50. http://www.guardian.co.uk/business/feedarticle/8021465.

51. http://uk.reuters.com/article/idUKLDE5BU0E020091231?pageNumber=1&virtualBrandChannel=0.

52. http://community.nytimes.com/comments/dotearth.blogs.nytimes.com/2009/07/07/more-ideas-for-breaking-climate-deadlock/?permid=1#comment1.

53. http://www.news-adhoc.com/gabriel-betont-notwendigkeit-neuer-kohlekraftwerke-idna2009031322122/.

54. http://www.bdtonline.com/local/local_story_136191739.html.

55. http://www.nature.com/nature/journal/v445/n7128/full/445595a.html; http://energycommerce.house.gov/Press_111/20090305/testimony_wara.pdf.

56. http://www.ft.com/cms/s/0/128a52de-deaf-11de-adff-00144feab49a.html.

57. http://www.news.com.au/story/0,27574,25863085-421,00.html.

58. See http://www.iop.org/EJ/article/1748-9326/4/3/034012/erl9_3_034012.html.

59. The totals from the doughnut graphs do not match up perfectly with the EIA data in Table 4.4 due to the use of different base years and rounding.

60. In Table 4.4 as well I ignore the realities of substituting electricity for liquid fuels. It is safe to conclude that the real-world challenge is even much greater than that implied by this exercise, which is sobering enough.

61. Or about 2 million solar thermal plants or 8 million wind turbines.

62. K. Caldeira, A. K. Jain, and M. I. Hoffert, "Climate Sensitivity Uncertainty and the Need for Energy Without CO_2 Emission," *Science* 299 (2003): 2052–2054.

Chapter 5

1. http://www.independent.co.uk/environment/climate-change/climate-scientists-its-time-for-plan-b-1221092.html.

2. http://www.ametsoc.org/POLICY/2009geoengineeringclimate_amsstatement.html.

3. For a very readable overview, see Eli Kintisch, *Hack the Planet* (New York: John Wiley and Sons, 2010).

4. http://news.bbc.co.uk/2/hi/7959570.stm.

5. Alvin Weinberg, "Can Technology Replace Social Engineering?" *Bulletin of the Atomic Scientists* (December 1966): 4–8.

6. http://dotearth.blogs.nytimes.com/2009/10/15/branson-on-space-climate-biofuel-elders/; http://blogs.ft.com/energy-source/2009/08/12/bjorn-lomborg-answers-readers-questions-on-geo-engineering-and-the-cost-of-avoiding-climate-change/.

7. http://www.theatlantic.com/issues/2000/07/sarewitz.htm.

8. http://www.nj.com/news/ledger/index.ssf?/base/news-11/117756219398850.xml&coll=1&thispage=1.

9. See Mike Hulme's excellent book, *Why We Disagree About Climate Change* (Cambridge: Cambridge University Press, 2009).

10. Weinberg, "Can Technology Replace Social Engineering?"

11. D. Sarewitz and R. Nelson, "Three Rules for Technological Fixes," *Nature* 456 (2008): 871–872.

12. I am grateful to Steve Rayner and Ben Hale for sharing their views on geoengineering, which has greatly shaped my thinking on this topic.

13. http://www.springerlink.com/content/t1vn75m458373h63/fulltext.pdf.

14. http://sciencepolicy.colorado.edu/prometheus/what-is-wrong-with-politically-motivated-research-3998.

15. http://royalsociety.org/Stop-emitting-CO2-or-geoengineering-could-be-our-only-hope/.

16. http://www.technologyreview.com/energy/24157/page2/.

17. M. Goes, K. Keller, and N. Tuana, "The Economics (or Lack Thereof) of Aerosol Geoengineering," *Climatic Change* (submitted).

18. http://www.publications.parliament.uk/pa/ld200910/ldhansrd/text/100114-0005.htm; http://www.theyworkforyou.com/lords/?id=2010-01-14a.617.6#g634.0.

19. National Academy of Science, Committee on Science Engineering and Public Policy, *Policy Implications of Greenhouse Warming: Mitigation, Adaptation, and the Science Base* (Washington, DC: National Academies Press, 1992).

20. http://www.imb.uq.edu.au/index.html?page=48437.

21. See James Scott, *Seeing Like a State* (New Haven: Yale University Press, 1998). A search of Google Scholar for "unintended consequences" and "complex systems" results in about 3,500 articles.

22. David R. Morrow et al., "Toward Ethical Norms and Institutions for Climate Engineering Research," *Environmental Research Letters* 4, no. 4 (2009).

23. Over 100 years, see http://www.ipcc.ch/publications_and_data/ar4/wg1/en/ch2s2-10-2.html.

24. Its large carbon dioxide equivalency also helps to explain why it is that the destruction of HFC-23 was so lucrative.

25. See, for example, Drew T. Shindell et al., "Improved Attribution of Climate Forcing to Emissions," *Science* 326, no. 5953 (2009): 716.

26. IPCC, "Working Group I: The Physical Science Basis of Climate Change," 2007, http://ipcc-wg1.ucar.edu/wg1/wg1-report.html.

27. National Research Council, Committee on Radiative Forcing Effects on Climate, *Radiative Forcing of Climate Change: Expanding the Concept and Addressing Uncertainties* (Washington, DC: National Academies Press, 2005).

28. http://www.thebulletin.org/files/064002006_0.pdf.

29. See the wide range of sources cited in National Research Council, Committee on Radiative Forcing Effects on Climate, *Radiative Forcing of Climate Change*.

30. http://www.springerlink.com/content/u60648p577607v87/.

31. http://americasclimatechoices.org/Geoengineering_Input/attachments/Travis_geoengineering_ACC.pdf.

32. Careful observers will note that this dynamic is similar to debates that already take place over extreme events and their relationship to carbon dioxide emissions.

33. http://blogs.wsj.com/environmentalcapital/2009/10/14/catch-me-if-you-can-does-the-ieas-carbon-capture-plan-make-any-sense/?utm_source=feedburner&utm_medium=feed&utm_campaign=Feedpercent3A+wsjpercent2Fenvironmentalcapitalpercent2Ffeed+percent28WSJ.compercent3A+Environmental+Capital+-+WSJ.compercent29.

34. http://www.newscientist.com/article/mg20126921.500-one-last-chance-to-save-mankind.html.

35. L. Orenstein et al., "Irrigated Afforestation of the Sahara and Australian Outback to End Global Warming," *Climatic Change* 97 (2009): 409–437.

36. http://www.nature.com/news/news-features/index.html. Eisenberger is former director of the Lamont-Doherty Earth Observatory at Columbia University and cofounder of the air-capture company Global Thermostat.

37. See R. A. Pielke, Jr., "An Idealized Assessment of the Economics of Air Capture of Carbon Dioxide in Mitigation Policy," *Environmental Science and Policy* 12, no. 3 (2009) 216–225.

38. R. Angamuthu et al., "Electrocatalytic CO_2 Conversion to Oxalate by a Copper Complex," *Science* 327 (2010): 313.

39. I further assume that the additional costs of storage are smaller than the uncertainties across estimates. See Pielke's "Idealized Assessment" for full details on a more rigorous idealized cost assessment.

40. These values are equivalent to $500, $360, and $100 per metric ton of carbon.

41. This calculation assumes a 2.5 percent 2010–2050 growth rate in global GDP. This is an average cost calculation. No effort has been made here to account for the time value of money or different approaches to calculating economic growth across countries, which have been discussed elsewhere in great depth in the context of climate change, and all dollars are expressed in constant-year terms.

42. N. Stern, *The Economics of Climate Change: The Stern Review* (Cambridge: Cambridge University Press, 2007). Relevant chapters are online at http://www.hmtreasury.gov.uk./media/F/0/Chapter_9_Identifying_the_Costs_of_Mitigation.pdf; and http://www.hm-treasury.gov.uk./media/B/7/Chapter_10_Macroeconomic_Models_of_Costs.pdf.

43. E. A. Parson, "Reflections on Air Capture: The Political Economy of Active Intervention in the Global Environment," *Climatic Change* 74 (2006): 5–15.

Chapter 6

1. http://www.ipcc.ch/pdf/assessment-report/ar4/wg1/ar4-wg1-spm.pdf.

2. James Hansen et al., "Climate Forcings in the Industrial Era," *Proceedings of the National Academy of Sciences* 95 (1998): 12753–12758; R. A. Pielke, Sr., "Overlooked Issues in the U.S. National Climate and IPCC Assessments," *Climatic Change* 52 (2002): 1–11.

3. The IPCC has been criticized as well for its emphasis on greenhouse gases at the expense of other important climate influences.

4. John W. Zillman, "The IPCC: A View from the Inside" (Australian APEC Study Centre, 1997), available at http://www.apec.org.au/docs/zillman.pdf. See also Zillman, "Climate Change," in *2003 Yearbook Australia* (Canberra: Australian Bureau of Statistics, 2003), 34–44.

5. See R. A. Pielke, Jr., "Rethinking the Role of Adaptation in Climate Policy," *Global Environmental Change* 8, no. 2 (1998): 159–170.

6. William Jefferson Clinton, "State of the Union Address" (2000), www.pub.whitehouse.gov/uri-res/I2R?pdi://oma.eop.gov.us/2000/01/27/15.text.1.

7. http://www.whitehouse.gov/the_press_office/Remarks-by-the-President-at-UN-Secretary-General-Ban-Ki-moons-Climate-Change-Summit.

8. R. A. Pielke, Jr., R. Klein, and D. Sarewitz, "Turning the Big Knob: Evaluating the Use of Energy Policy as a Means to Modulate Future Climate Impacts," *Energy and Environment* 11 (2000): 255–276.

9. Ernest Hollings, *Congressional Record* (1990), H17739; http://www.agu.org/meetings/fm09/lectures/lecture_videos/A23A.shtml.

10. See, for example, B. C. O'Neill and M. Oppenheimer, "Climate Change: Dangerous Climate Impacts and the Kyoto Protocol," *Science* 296 (2002): 1971–1972.

11. http://unfccc.int/resource/docs/convkp/conveng.pdf. The FCCC further states: "'Adverse effects of climate change' means changes in the physical environment or biota resulting from climate change which have significant deleterious effects on the composition, resilience or productivity of natural and managed ecosystems or on the operation of socio-economic systems or on human health and welfare."

12. http://www.whitehouse.gov/news/releases/2001/06/20010611-2.html; D. Bodansky, "U.S. Climate Policy After Kyoto: Elements for Success" (Carnegie Endowment for International Peace Policy Brief, 2002), available at http://www.ceip.org/files/pdf/Policybrief15.pdf.

13. There is a large literature on the notion of "dangerous interference." For example, O'Neill and Oppenheimer, "Climate Change"; S. Schneider, "What Is 'Dangerous' Climate Change?" *Nature* 411 (2001): 17–19; and M. Parry et al., "Millions at Risk: Defining Critical Climate Change Threats and Targets," *Global Environmental Change* 11 (2001): 181–183.

14. http://www.ipcc.ch/pub/un/syreng/spm.pdf.

15. S. Dessai et al., "Defining and Experiencing Dangerous Climate Change," *Climatic Change* 64 (2004): 11–25. More generally, on challenges in the management of highly complex systems, see L. Gunderson and C. S. Holling, *Panarchy: Understanding Transformations in Human and Natural Systems* (Washington, DC: Island Press, 2002); and J. X. Kasperson and R. E. Kasperson, *Global Environmental Risk* (Tokyo: United Nations University Press and Earthscan, 2001).

16. D. Sarewitz and R. A. Pielke, Jr., "Breaking the Global-Warming Gridlock," *Atlantic Monthly* 286, no. 1 (2000): 55–64.

17. R. B. Alley et al., "Abrupt Climate Change," *Science* 299 (2003): 2005–2010.

18. G. Marland et al., "The Climatic Impacts of Land Surface Change and Carbon Management, and the Implications for Climate-Change Mitigation Policy," *Climate Policy* 3 (2003): 149–157.

19. http://www.cop4.org/resource/docs/1996/sbsta/07a01.pdf.

20. An NRC report suggests that this key assumption may be incorrect. See National Research Council, Committee on Radiative Forcing Effects on Climate, *Radiative Forcing of Climate Change* (see chap. 5, n. 27), available at http://www.nap.edu/catalog/11175.html.

21. See, for example, the controversy over chapter 8 (on detection and attribution) of the IPCC Second Assessment report: M. Lahsen, "The Detection and Attribution of Conspiracies: The Controversy over Chapter 8," in *Paranoia Within Reason: A Casebook on Conspiracy as Explanation*, vol. 6, Late Editions Series of Cultural Studies for the End of the Century (Chicago: University of Chicago Press, 1999).

22. Geoengineering, as discussed in the previous chapter, is a more recent addition to the list of policy options.

23. A comprehensive treatment of the inconsistency in definitions of "climate change" goes beyond the present scope but is a subject worth further investigation. See, for example, E. Larsson, "Science and Policy in the International Framing of the Climate Change Issue" (master's thesis, Linköpings Universitet, 2004).

24. IPCC Second Assessment Synthesis of Scientific-Technical Information relevant to interpreting Article 2 of the UN Framework Convention on Climate Change, http://www.unep.ch/ipcc/pub/sarsyn.htm.

25. Pielke and Sarewitz, "Breaking the Global-Warming Gridlock."

26. Discussed in R. A. Pielke, Jr., "Misdefining 'Climate Change': Consequences for Science and Action," *Environmental Science and Policy* 8 (2005): 548–561.

27. P. Brown, "Russia Urged to Rescue Kyoto Pact," *The Guardian*, February 26, 2003; UNEP press release, "Global Warming: Asia Vulnerable," February 20, 2001, http://206.67.58.208/uneproap/html/nr/nr01-01.htm.

28. Brown, "Russia Urged to Rescue Kyoto Pact."

29. R. J. Lempert and M. E. Schlesinger, "Robust Strategies for Abating Climate Change," *Climatic Change* 45 (2000): 387–401.

30. http://www.ipcc.ch/pub/un/syreng/spm.pdf.

31. The notion that there may be both winners and losers under climate change is not new. See, for example, Mickey H. Glantz, "Assessing the Impacts of Climate: The Issue of Winners and Losers in a Global Climate Change Context," in *Climate Change Research: Evaluation and Policy Implications*, edited by S. Zwerver et al. (Amsterdam: Elsevier, 1995), 41–54, quotes on 44.

32. D. Sarewitz, "How Science Makes Environmental Controversies Worse," *Environmental Science and Policy* 7 (2004): 385–403.

33. See, for instance, G. Prins et al., *How to Get Climate Policy Back on Course* (London: Institute for Science, Innovation, and Society, Oxford University and London School of Economics, the Mackinder Programme, 2009).

34. The climate community already recognizes this need through an odd bit of jargon that describes the "mainstreaming" of adaptation, which means bringing adaptation more in line with conventional discussions of development.

35. http://www.nature.com/climate/2010/1002/full/climate.2010.06.html.

36. See Sarewitz, Pielke, and Byerly, *Prediction* (see chap. 1, n. 55).

37. R. A. Pielke, Jr., and D. Sarewitz, "Wanted: Scientific Leadership on Climate," *Issues in Science and Technology* (Winter 2003): 27–30.

Chapter 7

1. Since 1980 damage from earthquake-related disasters has increased at a rate less than that due to weather-related disasters. Some claim naively that this

indicates a signal of greenhouse gas emissions in weather-related disasters. As we will see, this is not the case.

2. I thank Laurens Bouwer for sharing a prepublication of a review paper that systematically looks at this literature.

3. C. W. Landsea et al., "Downward Trends in the Frequency of Intense Atlantic Hurricanes During the Past Five Decades," *Geophysical Research Letters* 23 (1996): 1697–1700.

4. R. A. Pielke, Jr., and C. W. Landsea, "Normalized Hurricane Damages in the United States, 1925–95," *Weather and Forecasting* (American Meteorological Society) 13 (1998): 621–631.

5. R. A. Pielke, Jr., et al., "Normalized Hurricane Damages in the United States, 1900–2005," *Natural Hazards Review* 9, no. 1 (2008): 29–42. Since then we have developed a Web site that allows one to examine the loss data updated through 2009 in the context of contemporary storms: http://www.icatdamageestimator.com.

6. In fact, U.S. hurricane losses are the dominant factor within the Munich Reinsurance global loss data set.

7. See, for instance, C. W. Landsea, "Hurricanes and Global Warming," *Nature* 438 (2005), available at doi:10.1038/nature04477; and R. L. Smith, "Statistical Trend Analysis in Weather and Climate Extremes in a Changing Climate: Regions of Focus—North America, Hawaii, Caribbean, and U.S. Pacific Islands," in *A Report by the U.S. Climate Change Science Program and the Subcommittee on Global Change Research*, edited by T. R. Karl et al. (Washington, DC: U.S. Climate Change Science Program, 2008).

8. See R. A. Pielke, Jr., "United States Hurricane Landfalls and Damages: Can One- to Five-Year Predictions Beat Climatology?" *Environmental Hazards* 8 (2009): 187–200; and R. A. Pielke, Jr., and C. W. Landsea, "La Niña, El Niño, and Atlantic Hurricane Damages in the United States," *Bulletin of the American Meteorological Society* 80, no. 10 (1999): 2027–2033.

9. http://sciencepolicy.colorado.edu/prometheus/replications-of-our-normalized-hurricane-damage-work-4480.

10. http://www.nature.com/ngeo/journal/vaop/ncurrent/pdf/ngeo779.pdf.

11. R. A. Pielke, Jr., "Future Economic Damage from Tropical Cyclones: Sensitivities to Societal and Climate Changes," *Philosophical Transactions of the Royal Society* 365, no. 1860 (2007): 1–13.

12. IDAG [International Ad Hoc Detection and Attribution Group], "Detecting and Attributing External Influences on the Climate System: A Review of Recent Advances," *Journal of Climate* 18 (2005): 1291–1314.

13. http://www.springerlink.com/content/m0w7058402331283/.

14. See, for example, http://www.sciencemag.org/cgi/content/full/313/5789/940.

15. http://ams.allenpress.com/perlserv/?request=get-abstract&doi=10.1175%2F2009JCLI2683.1.

16. http://www.sciencemag.org/cgi/content/short/327/5964/454.

17. http://www.who.int/whr/2002/en/summary_riskfactors_chp4.pdf, 26.

18. http://www.who.int/publications/cra/chapters/volume2/1543-1650.pdf.

19. http://www.who.int/publications/cra/chapters/volume2/1543-1650.pdf, 1544–1545.

20. http://www.who.int/healthinfo/global_burden_disease/estimates_country/en/index.html.

21. http://www.grida.no/climate/ipcc_tar/wg2/321.htm.

22. http://www.grida.no/climate/ipcc_tar/wg2/325.htm.

23. Munich Reinsurance Group, *Topics, 2000: Natural Catastrophes, the Current Position* (Munich: Münchener Rückversicherung-Gesellschaft, 2000), available at http://www.munichre.com/publications/302-02354_en.pdf?rdm=80335.

24. Mills's commentary is a fact checker's nightmare, because it misreferences not just the IPCC but an entire bibliography's worth of analysis, as documented in detail at http://sciencepolicy.colorado.edu/prometheus/the-other-hockey-stick-3566.

25. http://sciencepolicy.colorado.edu/sparc/research/projects/extreme_events/munich_workshop/muirwood.pdf.

26. S. Miller, R. Muir-Wood, and A. Boissonnade, "An Exploration of Trends in Normalized Weather-Related Catastrophe Losses," in *Climate Extremes and Society*, edited by H. F. Diaz and R. J. Murnane, 225–247 (Cambridge: Cambridge University Press, 2008).

27. Much of this literature can be found cited in the background papers to the Hohenkammer Workshop report, http://sciencepolicy.colorado.edu/sparc/research/projects/extreme_events/munich_workshop/workshop_report.html.

28. http://www.timesonline.co.uk/tol/news/environment/article7009710.ece.

29. My analysis appeared as a peer-reviewed paper: R. A. Pielke, Jr., "Mistreatment of the Economic Impacts of Extreme Events in the 'Stern Review Report' on the Economics of Climate Change," *Global Environmental Change* 17 (2007): 302–310.

30. The manner in which Stern calculated the costs of climate change did not involve the disaster cost estimates. I arrive at the 40 percent figure by comparing the lower bound of Stern's estimates of the costs of climate change, 5 percent, with an estimate that extreme events cost several percent of GDP: 2/5 = 40 percent. Richard Tol says of this situation, "As you have shown . . . Stern's numbers about extreme weather are nonsense. As I and others have shown . . . Stern's total impact estimates are nonsense too. What share does one nonsense contribute to another nonsense?"

31. http://www.ipcc.ch/pdf/assessment-report/ar4/wg2/ar4-wg2-chapter1.pdf, 110. Robert Muir-Wood, of Risk Management Solutions, was an author of the chapter of the IPCC report that selectively highlighted his own non-peer-reviewed work.

32. http://www.ipcc.ch/pdf/assessment-report/ar4/wg2/ar4-wg2-chapter1sm.pdf. The figure was produced by Muir-Wood and never appeared in any other literature besides the IPCC report, peer reviewed or otherwise.

33. This is confirmed in a press release issued by Muir-Wood's employer, Risk Management Solutions, which explains why the graph was miscited by the IPCC: http://www.rms.com/Publications/2010_FAQ_IPCC.pdf.

34. Miller, Muir-Wood, and Boissonnade, "Exploration of Trends."

35. http://rogerpielkejr.blogspot.com/2010/01/what-does-pielke-think-about-this.html.

36. R. A. Pielke, Jr., "Seventh Annual Roger Revelle Commemorative Lecture: Disasters, Death, and Destruction—Making Sense of Recent Calamities," *Oceanography*, special issue, *Oceans and Human Health* 19, no. 2 (2006): 138–147.

37. H. E. Brooks and C. A. Doswell III, "Normalized Damage from Major Tornadoes in the United States, 1890–1999," *Weather and Forecasting* (American Meteorological Society) 16 (2001): 168–176; Pielke et al., "Normalized Hurricane Damages in the United States, 1900–2005"; M. Downton, J. Z. B. Miller, and R. A. Pielke, Jr., "Reanalysis of U.S. National Weather Service Flood Loss Database," *Natural Hazards Review* 6 (2005): 13–22.

38. http://e360.yale.edu/content/feature.msp?id=2245.

39. http://discovermagazine.com/2010/apr/10-it.s-gettin-hot-in-here-big-battle-over-climate-science/article_view?searchterm=michael%20mann&b_start:int=0.

40. http://www.google.com/hostednews/ap/article/ALeqM5h15IYOYm3RQQ9sySx-w-c1dYWJ2wD9EANS781.

41. http://www.economist.com/world/international/displaystory.cfm?story_id=12208005.

Chapter 8

1. http://www.monbiot.com/archives/2009/11/23/the-knights-carbonic/.

2. Indeed, this was confirmed by an independent investigation that reported in April 2010: http://www.uea.ac.uk/mac/comm/media/press/CRUstatements/Report+of+the+Science+Assessment+Panel.

3. http://www.ft.com/cms/s/0/cc90fb80-e817–11de-8a02–00144feab49a.html.

4. http://www.scpr.org/news/2009/11/25/stolen-emails-raise-questions-on-climate-research/.

5. http://pielkeclimatesci.wordpress.com/2009/11/24/beware-saviors-by-demetris-koutsoyiannis/.

6. Citizens of other countries are also identified as not understanding the science. See, for example, http://www.telegraph.co.uk/earth/earthnews/6253912/Most-people-in-denial-over-climate-change-according-to-psychologists.html.

7. http://www.lowyinstitute.org/PublicationPop.asp?pid=1167.

8. http://www.guardian.co.uk/environment/2009/sep/28/us-climate-change-copenhagen-schellnhuber.

9. http://www.telegraph.co.uk/earth/earthnews/6253912/Most-people-in-denial-over-climate-change-according-to-psychologists.html.

10. http://correspondents.theatlantic.com/conor_clarke/2009/07/an_interview_with_thomas_schelling_part_two.php.

11. http://extras.timesonline.co.uk/pdfs/sjp_memorandum_290509.pdf. "Stimulated by the manifesto of Bertrand Russell and Albert Einstein, the first Pugwash gathering of 1957 united scientists of all political persuasions to discuss the threat posed to civilization by the advent of thermonuclear weapons. Global climate change represents a threat of similar proportions, and should be addressed in a similar manner."

12. http://www.bbc.co.uk/blogs/climatechange/2009/05/climate_change_versus_nuclear_armageddon.html.

13. http://www.gallup.com/poll/116590/Increased-Number-Think-Global-Warming-Exaggerated.aspx. It is likely that the events following the release of the East Anglia e-mails will exacerbate this trend.

14. http://www.guardian.co.uk/environment/2009/feb/11/climate-change-science-pope.

15. http://www.guardian.co.uk/environment/2009/mar/11/amazon-global-warming-trees.

16. http://www.guardian.co.uk/commentisfree/2009/apr/07/amazon-rainforest-global-warming.

17. http://iopscience.iop.org/1748–9326/4/2/020201/pdf/1748–9326_4_2_020201.pdf.

18. http://www.washingtonpost.com/wp-dyn/content/article/2009/12/04/AR2009120404511.html.

19. Stephen H. Schneider, "Don't Bet All Environmental Changes Will Be Beneficial," APS News Online, August–September 1996.

20. Ibid.

21. http://www.newyorker.com/archive/2006/11/20/061120fa_fact_kolbert. When I pointed out this statement on my blog at the time, Caldeira responded that this statement had been made up by Kolbert. Elizabeth Kolbert did not respond to my queries on this, and the *New Yorker* never addressed the apparent dispute. Like much in the climate debate, this is a case of she said/he said.

22. http://seedmagazine.com/content/article/is_there_a_better_word_for_doom/.

23. http://w3g.gkss.de/staff/storch/Media/climate.culture.041130.pdf.

24. http://sciencepolicy.colorado.edu/zine/archives/1–29/txt/zine28.txt.

25. http://www.americanscientist.org/issues/pub/liberating-science-from-politics/1.

26. http://online.wsj.com/article/SB10001424052748704107104574571613215771336.html.

27. http://rogerpielkejr.blogspot.com/2010/01/sorry-but-this-stinks.html.

28. http://www.dailymail.co.uk/news/article-1245636/Glacier-scientists-says-knew-data-verified.html. The journalist who wrote this story explained to me in an interview that he adamantly stands behind what he reported, despite the scientist's later claims that the quote was invented. The allegation by a blogger of the scientist being misquoted can be found here: http://climateprogress.org/2010/01/25/un-scientist-refutes-daily-mail-claim-himalayan-glacier-2035-ipcc-mistake-not-politically-motivated/.

29. Rayner is a close colleague of mine. His complaint can be found at http://sciencepolicy.colorado.edu/prometheus/archives/EAC%20memo%20fin.doc.

30. http://www.economist.com/agenda/displayStory.cfm?story_id=3630425.

31. http://www.slate.com/id/2248236/pagenum/all.

32. http://pewresearch.org/pubs/1276/science-survey.

33. http://www.nytimes.com/2008/08/01/opinion/01krugman.html.

34. http://rogerpielkejr.blogspot.com/2009/12/climate-scientist-threatens-boycott-of.html.

35. http://www.slate.com/id/2247487/pagenum/all/.

36. http://blog.algore.com/2007/03/als_testimony_before_the_house.html.

37. S. Jasanoff, *The Fifth Branch: Science Advisors as Policymakers* (Cambridge: Harvard University Press, 1990); http://www.slate.com/id/2247487/.

Chapter 9

1. This number is surely an underestimate, as the figure of 4.4 metric tons per person includes those who lack energy access. See http://www.worldenergyoutlook.org/docs/weo2009/climate_change_excerpt.pdf.

2. http://www.johnkay.com/2004/01/17/obliquity/; J. Kay, *Obliquity: Why Our Goals Are Best Achieved Indirectly* (London: Profile Books, 2010).

3. http://www.google.org/rec.html.

4. http://www.google.org/RE-C_Brief.pdf.

5. http://ideas.repec.org/a/ces/ifofor/v10y2009i3p10–13.html.

6. http://www.cspo.org/projects/eisbu/report.pdf.

7. http://www.cgiar.org/.

8. http://www.gavialliance.org/.

9. http://fixtheclimate.com/fileadmin/templates/page/scripts/downloadpdf.php?file=/uploads/tx_templavoila/AP_Technology_Galiana_Green_v.6.0.pdf.

10. http://www.newyorker.com/online/blogs/stevecoll/2009/01/carbon-taxes-vs.html.

11. http://indiabudget.nic.in/ub2010–11/bs/speecha.htm.

12. See Shellenberger et al. for a largely consistent set of proposals: http://thebreakthrough.org/blog/Fast%20Clean%20Cheap.pdf.

13. http://www.eia.doe.gov/emeu/international/contents.html.

14. http://www.bostonherald.com/news/us_politics/view.bg?articleid=1239353.

15. http://www.washingtonpost.com/wp-dyn/content/article/2010/03/21/AR2010032101711.html.

16. http://news.discovery.com/earth/whales-carbon-climate-change.html.

Index